Mathematical Problem Factories

Almost Endless Problem Generation

Synthesis Lectures on Mathematics and Statistics

Editor
Steven G. Krantz, *Washington University, St. Louis*

Analytical Techniques for Solving Nonlinear Partial Differential Equations
Daniel J. Arrigo
2019

Aspects of Differential Geometry IV
Esteban Calviño-Louzao, Eduardo García-Río, Peter Gilkey, JeongHyeong Park, and Ramón Vázquez-Lorenzo
2019

Symmetry Problems. The Navier–Stokes Problem.
Alexander G. Ramm
2019

An Introduction to Partial Differential Equations
Daniel J. Arrigo
2017

Numerical Integration of Space Fractional Partial Differential Equations: Vol 2 - Applications from Classical Integer PDEs
Younes Salehi and William E. Schiesser
2017

Numerical Integration of Space Fractional Partial Differential Equations: Vol 1 - Introduction to Algorithms and Computer Coding in R
Younes Salehi and William E. Schiesser
2017

Aspects of Differential Geometry III
Esteban Calviño-Louzao, Eduardo García-Río, Peter Gilkey, JeongHyeong Park, and Ramón Vázquez-Lorenzo
2017

The Fundamentals of Analysis for Talented Freshmen
Peter M. Luthy, Guido L. Weiss, and Steven S. Xiao
2016

Aspects of Differential Geometry II
Peter Gilkey, JeongHyeong Park, and Ramón Vázquez-Lorenzo
2015

Aspects of Differential Geometry I
Peter Gilkey, JeongHyeong Park, and Ramón Vázquez-Lorenzo
2015

An Easy Path to Convex Analysis and Applications
Boris S. Mordukhovich and Nguyen Mau Nam
2013

Analytical Techniques for Solving Nonlinear Partial Differential Equations
Daniel J. Arrigo
2019

Aspects of Differential Geometry IV
Esteban Calviño-Louzao, Eduardo García-Río, Peter Gilkey, JeongHyeong Park, and Ramón Vázquez-Lorenzo
2019

Symmetry Problems. The Navier–Stokes Problem.
Alexander G. Ramm
2019

An Introduction to Partial Differential Equations
Daniel J. Arrigo
2017

Numerical Integration of Space Fractional Partial Differential Equations: Vol 2 - Applications from Classical Integer PDEs
Younes Salehi and William E. Schiesser
2017

Numerical Integration of Space Fractional Partial Differential Equations: Vol 1 - Introduction to Algorithms and Computer Coding in R
Younes Salehi and William E. Schiesser
2017

Aspects of Differential Geometry III
Esteban Calviño-Louzao, Eduardo García-Río, Peter Gilkey, JeongHyeong Park, and Ramón Vázquez-Lorenzo
2017

The Fundamentals of Analysis for Talented Freshmen
Peter M. Luthy, Guido L. Weiss, and Steven S. Xiao
2016

Aspects of Differential Geometry II
Peter Gilkey, JeongHyeong Park, and Ramón Vázquez-Lorenzo
2015

Aspects of Differential Geometry I
Peter Gilkey, JeongHyeong Park, and Ramón Vázquez-Lorenzo
2015

An Easy Path to Convex Analysis and Applications
Boris S. Mordukhovich and Nguyen Mau Nam
2013

A Gyrovector Space Approach to Hyperbolic Geometry
Abraham Albert Ungar
2008

Mathematical Problem Factories: Almost Endless Problem Generation

Andrew McEachern and Daniel Ashlock

ISBN: 978-3-031-01308-9   paperback
ISBN: 978-3-031-02436-8   ebook
ISBN: 978-3-031-00282-3   hardcover

DOI 10.1007/978-3-031-02436-8

A Publication in theSpringer series
SYNTHESIS LECTURES ON MATHEMATICS AND STATISTICS

Lecture #44
Series Editor: Steven G. Krantz, *Washington University, St. Louis*
Series ISSN
Print 1938-1743   Electronic 1938-1751

Mathematical Problem Factories

Almost Endless Problem Generation

Andrew McEachern
York University

Daniel Ashlock
University of Guelph

SYNTHESIS LECTURES ON MATHEMATICS AND STATISTICS #44

ABSTRACT

A *problem factory* consists of a traditional mathematical analysis of a type of problem that describes many, ideally all, ways that the problems of that type can be cast in a fashion that allows teachers or parents to generate problems for enrichment exercises, tests, and classwork. Some problem factories are easier than others for a teacher or parent to apply, so we also include banks of example problems for users. This text goes through the definition of a problem factory in detail and works through many examples of problem factories. It gives banks of questions generated using each of the examples of problem factories, both the easy ones and the hard ones. This text looks at sequence extension problems (what number comes next?), basic analytic geometry, problems on whole numbers, diagrammatic representations of systems of equations, domino tiling puzzles, and puzzles based on combinatorial graphs. The final chapter previews other possible problem factories.

KEYWORDS

problem generation, math problems, math education, problem factories

Contents

Preface

This book started on the way home from the Canadian Mathematical Education Study Group at Queens University in Kingston. Dr. McEachern was an organizer of the conference and Dr. Ashlock attended at Dr. McEachern's invitation. Driving home from Kingston, Ontario to Guelph, Ontario is a long trip and a number of interesting problems for students were presented at the conference. A long car trip is a fertile environment to think about things. Several of the problems from the conference generalized nicely into families of problems. These generalizations took the form of proving modest theorems that opened up whole spaces of problems. These spaces are often infinite, but human beings are not, which is why the title of this book is *Almost* Endless Problem Generation: only a large, finite number of the problems that arise are typically suitable to be posed to humans.

We decided to call the problem generation methods we found *problem factories*. Problem factories should not be too difficult to follow, particularly the practical directions for making problems that are the result of the initial mathematical investigation. Some of the problem factories are our original creations, others we are pulling in, generalizing, and reformatting. One of our problem factories generalizes a problem that appeared in one of the *Die Hard* movies and another generalizes problems appearing in a wonderful collection of problems called *Bovine Math* by Professor Peter Harrison of Queens University.

Many excellent math problems do not fit naturally into the problem factory framework. If, after reading Chapter 1, you decide to make your own problem factories (which we heartily encourage), keep this in mind. The generalization of a killer problem must be natural, it must *flow* from the mathematics. We hope the problem factories we chose for this volume meet that standard and we hope you find the book both useful and entertaining.

Andrew McEachern and Daniel Ashlock
August 2021

Acknowledgments

The authors would like to thank their partners for helping create an environment that made it possible to write this book and the many students that stimulated us to realize the book might be useful, but especially we thank Professor Peter Harrison for giving us the example that snapped the entire theme of the book into place.

Andrew McEachern and Daniel Ashlock
August 2021

CHAPTER 1

What Are Problem Factories?

A *problem factory* is an abstract math problem or a class of abstract math problems that can be solved, in general, possibly using advanced techniques, while yielding a supply of problems or puzzles accesible to K–12 or early undergraduate students. This book gives examples of solved problem factories and the problems that arise from them. Typically, a problem factory starts with a good problem and generalizes it to a family of problems. A problem that initiates a problem factory is ideally one that is moderately difficult and requires a leap of intuition or a clever idea to solve.

A rigorous definition of a problem factory is not available and would probably be more limiting than helpful. This book serves as a collection of examples and, as we will see, there is a spectrum of both difficulty and ingenuity in problem factories. A problem factory should be able to generate a large number of problems and a good problem factory will generate a large number of problems accessible to most students at its level. The problem factories in this book are accompanied by collections of workable example problems.

One reason to create problem factories is to permit a teacher to have a variety of problems with a similar character but different details, a form of question bank that can be run without a computer, though they can be put into computerized repositories as well. Students can Google problems and join "tutoring" websites that keep their own banks of problems, so being able to smoothly change the details of a problem and, in many cases, tune the difficulty of problems, in the form of problem factories will allow a teacher a form of countermeasure. An early title for this work that was discarded was "academic countermeasures."

Problem factories can also be used as longer-term exercises. Giving a group of students a collection of problems from the same problem factory leaves room for the students to notice patterns, develop general solution methods, and even deduce the ideas underlying the problem factory. We now dive right in with some examples of problem factories appearing in this book.

1.0.1 SIMPLE ANALYTIC GEOMETRY PROBLEMS

Motivating problem: Find a square in the plane whose corners are points with whole numbers coordinates so that the square has an area of 5 square units. A solution is shown in Figure 1.1; note translations, reflections, and rotations of this solution—that keep integer coordinates at the corners—also work.

This problem is an example of a problem that leads naturally to a problem factory. There is a tendency to think of squares as lining up with the x- and y-axes. If a square of area 5 is

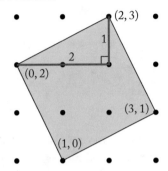

Figure 1.1: An example of a square with area 5 with corners with whole number coordinates.

lined up with the coordinate axes then the sides must have a length of $\sqrt{5}$ which makes points with whole number coordinates impossible. Students will tend to try, initially, to find a square aligned with the coordinate axes. The square that solves the problem must have sides of length $\sqrt{5}$—but it does this by exploiting the Pythagorean theorem which gives us a length of $\sqrt{5}$ via $1^2 + 2^2 = \sqrt{5}^2$. The intuitive leap required is to relax the unneeded assumption of alignment with the coordinate axes.

This problem inspires a problem factory in a straightforward fashion. If c is not a perfect square and $a^2 + b^2 = c$ then we can find a square—using the same sort of tilted square setup that solves the area 5—that has an area of c and which cannot align with the coordinate axes. If c is a perfect square then a solution aligned with the coordinate axes is available—which is a much easier problem. It might be good to put one of these easier problems into a group of these problems. This family of problems is developed at greater length as part of Chapter 3.

1.0.2   THE "DIE HARD" PROBLEM

Motivating problem: The problem from the movie *Die Hard* is as follows. There is a bomb is in a briefcase with an electronic scale. The characters have a 5-gallon jug and a 3-gallon jug. They are standing next to a fountain where they can take as much water as they want. They have 5 minutes to put one of the jugs on the scale with exactly 4 gallons of water in it, or the bomb will detonate.

The solution is not too hard. Fill the five gallon jug. Use the five gallon jug to fill the three gallon jug, leaving two gallons behind. Empty the three gallon jug. Pour the two gallons from the five gallon jug into the three gallon jug. Fill the five gallon jug again. Use the five gallon jug to top off the three gallon jug—which already has two gallons in it. That means that four gallons remain in the five gallon jug, problem solved.

Turning this into a problem factory is a little harder—if you just make up some jug sizes and a final amount, it is very likely you will pose a problem that has no solutions. They key to the problem factory is this: you can measure out any amount that is a multiple of the least common divisor of the sizes of the jugs. The *Die Hard* problem can be solved without needing another container—but some jug problems cannot be solved without another container to hold intermediate results. You might, for example, need a large basin or bathtub to measure out a large multiple of the least common multiple of the jug sizes. This problem factory is explored in detail in Chapter 4.

1.0.3   WHAT IS THE NEXT NUMBER?

Motivating problem: Find the next number in the sequence:

$$1, 4, 7, 10, 13, 16, ?$$

This sequence has next number 19 because the sequence follows the very simple pattern $3n + 1$ for $n = 0, 1, 2, \ldots$. Starting with one, keep adding three to get the next term. It is fairly obvious how sequence problems of this sort are a problem factory—just change the numbers "3" and "1" into something else. Just doing that, however, misses the fact that there are many possible patterns or templates for generating sequences. There is also another problem.

One of the authors was privileged to witness a lecture at Caltech by Douglas Hofstadter. Dr. Hofstadter was attempting to make a point about the nature of intelligence and used a sequence puzzle as an example. He stated that the next number in the sequence was obvious. The audience enthusiastically rebutted Dr. Hofstadter's claim by finding six defensible answers for the question "what is the next number?" The ensuing vigorous exchange of views exemplifies the problem that sequences following even a moderately complex pattern might fit several patterns.

This is a good example of a situation where a Ph.D. can be a handicap. Usually a sequence has a fairly obvious, fairly simple pattern. A repeat of the "Hofstadter incident" can be avoided by reserving some additional terms of the sequence and releasing them, if a pattern other than the one intended is discovered. We feel that if the students give a next number with a defensible rule, then they have solved the problem. Chapter 1 is about the problem factories that employ sequence patterns. The chapter attempts to assign relative hardness to different sequence templates, including some derived from as yet unsolved mathematical problems, like the *hailstone sequence*. There is no need to solve these unsolved problems to solve the sequence patterns—rather these unsolved problems give rise to pleasantly challenging sequence completion problems.

1.0.4   DIAGRAMMATIC LINEAR SYSTEMS

Motivating Problem: Given examples of bouquets of flowers with prices, can you deduce the price of the last bouquet? You may assume that the price of individual flowers is constant across all the bouquets.

Consider the problem posed by the images in Figure 1.2.

If we sum the two diagrams with known prices, this tells us that three of each sort of flower together costs $18.00 and so the cost of one flower of each type together is $6.00. Adding one

| $10.00 | $8.00 | $??.?? |

Figure 1.2: An example of a diagrammatic linear system made of flowers.

flower of each type to the first priced example gives us the correct number of flowers for the target configuration and its price: $16.00.

The mathematical formalism for this problem is systems of equations—a linear solver could be used to find the price of each flower (if the problem is well-posed) and then recover the cost of the target assortment of flowers. Many linear systems, however, can be solved with fewer steps that those required for a full linear system solution. It is also possible to pose problems where the prices of individual flowers cannot be deduced, but the price of the final bouquet can still be deduced. The technique given above, manipulation of diagrams, is such a special purpose short-cut from the point of view of solving linear systems—but it is also an effective warm-up exercise for solving linear systems.

Chapter 5 covers diagrammatic linear systems, giving a problem factory for generating these problems. There are many ways to reskin these problems, using items other than flowers and they can be tuned to having many different levels of difficulty. We reskin these problems using collectable cards and chocolates. The problem posed by one set of examples, such as the first two bouquets in Figure 1.2, can be used to pose multiple problems in the form of additional bouquets with unknown prices.

1.0.5 POLYOMINO TILING PUZZLES

A *polyomino* is a contiguous shape made of squares joined on the sides of the squares. The domino, or rectangular 2-omino, is the one with which most people are familiar. There are a variety of types of problems that can be posed about polyominos.

Motivating problem: Fill a 5×9 rectangle with three-square L-shaped polyominos. Figure 1.3 shows an example of tiling a rectangle with a simple polyomino. One of the problem factories in Chapter 6 concerns which sizes of rectangles can be tiled with a given polyomino. Once you

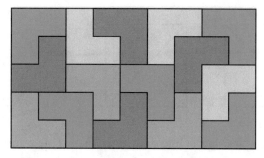

Figure 1.3: An example of tiling a 5 × 9 rectangle with the L-shaped 3-omino.

Figure 1.4: This is a tiling of a 6 × 5 rectangle with dominoes. It has the interesting property that the tiling does not split the original 6 × 5 rectangle into two rectangles along any vertical or horizontal lines.

know which sizes of rectangles can be filled with a polyomino the "fill the rectangle" problems become an infinite collection of puzzles.

Some polyominos cannot tile rectangles at all—their shape prevents it—so this class of problem factories includes unsolvable problems, a pedagogically useful commodity if used responsibly and sparingly. For a given polyomino the *spectrum* of that polyomino is the set of sizes of rectangles that can be tiled by the polyomino. Finding the spectra of different polyominos is a natural problem factory, but one that poses substantially harder problems.

There are more subtle problems that can be posed based on polyomino tiling, even with relatively simple polyominos like the domino. Look at the domino tiling of a 6 × 5 rectangle shown in Figure 1.4. This is the smallest possible domino tiling of a rectangle that does not split the original 6 × 5 rectangle into two rectangles along a vertical or horizontal line.

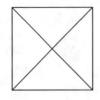

Figure 1.5: These are examples simple line drawings. The problem factory that generates these problems tells us by a simple test that the first and third can be drawn as a continuous line, but the second cannot.

Finding indecomposable tilings (meaning you can't split an original rectangle into smaller rectangles along a vertical or horizontal line) and tilings with related properties form a problem factory, and there are several interesting problems that are related to this problem factory, e.g., show that the tiling in Figure 1.4 are a small as possible.

1.0.6   CAN YOU DRAW THIS WITHOUT LIFTING YOUR PENCIL?

Motivating problem: Asking if a line drawing can be reproduced as a continuous line is a question based on the classical problem of finding Euler cycles in a graph. Examples of these sorts of problems appear in Figure 1.5.

The Euler cycle theorem is a part of *graph theory*, an area of math which has proven, thus far, to be rich in problem factories. Graph theory is also another source of unintuitively impossible problems. One of these is the famous *utility problem*, posed in Figure 1.6.

Connecting three objects to three others without a crossing is known to be impossible, there are proofs of this in most texts on graph theory. The utility problem is one way of skinning this fact to make an interesting problem. Chapter 7 covers problems that arise from graph theory. With this brief introduction to the art of the problem factory, we are ready to get more specific, moving on to the individual chapters.

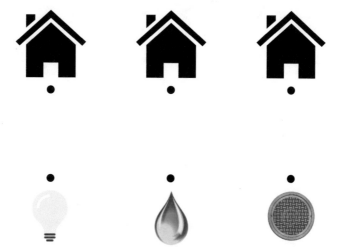

Figure 1.6: The utility problem: can three houses be connected to water, electricity, and sewage without any of the connections crossing one another?

CHAPTER 2

Sequence Extension Problems

Problem factories for sequence extension problems have at their core a list of generic sequence patterns. These patterns range from simple arithmetic sequences to more complex recursions and sequences that merge other sequences to form complex patterns. The sequences in this section use only whole numbers, a constraint that can be relaxed to generate more problems. Example 2.1 shows what a typical problem looks like.

The chapter is organized into a collection of short sections on various sequence patterns followed by a large block of example problems taken from all of the patterns so that, if you need a problem in a hurry, the ready-to-use problems are all gathered together in one place.

Example 2.1 What comes next?

$$3, 5, 9, 17, 33, ?$$

Answer: 65; one possible rule is to double the last term and subtract one.

Example 2.1 is simple, but would be even simpler if we made the numbers smaller. The simplicity, however, primarily resides in the fact that the sequence arises from a simple linear function. Simple linear functions are our first type of sequence problem generator.

A note on sequence notation. When we write s_n, we are referring to a sequence with a list of n numbers in that list. This list could be of infinite length. When we write s_1, we are usually referring to the first number in the list, generated by substituting $n = 1$ into the formula that is equal to s_n.

2.1   SIMPLE LINEAR SEQUENCE GENERATORS

A simple linear rule has the form

$$s_n = a \cdot n + b$$

for whole numbers a and b. Factors that control hardness of these problems are as follows.

- Small a make easy patterns, larger a make harder patterns, but past a certain point making a really large does not increase difficulty much.

- Setting $b = 0$ makes a much easier sequence, choosing a b that has factors in common with a sometimes increases the difficulty of the problem a little, and making b negative buys you some additional difficulty.

- Sequences with large b and negative a count down, which can make a problem more challenging.

2.1.1   SOLUTION METHODS

Something that will give you traction on a number of sequence guessing problems is computing the difference between adjacent terms, sometimes referred to as finite differences:

$$[a(n + 1) + b] - [an + b] = a(n + 1 - n) + (b - b) = a$$

and once you have a it is more-or-less game over for a linear rule problem.

Example 2.2   Find the next number: 6, 15, 24, 33, 42, ?

Solution: $15 - 6 = 9$; $24 - 15 = 9$; $33 - 24 = 9$; $42 - 33 = 9$; so we deduce that this sequence grows by 9 each time. The next number is 51.

Notice that, in solving Example 2.2, we did not need to deduce the full rule $s_n = an + b$. Knowledge of a sufficed to solve the problem. A slightly harder problem would be to ask the students to find the rule for the sequence.

2.2   QUADRATIC SEQUENCE GENERATORS

Next up after linear rules are the quadratic rules

$$s_n = a \cdot n^2 + b \cdot n + c$$

for whole numbers a, b, and c. Factors that control hardness of these problems are as follows.

- Making b or c equal to zero make the problem easier.

- Letting c be larger than about 4 make the problem quite a bit harder.

- Varying the signs of b and c increases hardness. If you make one positive, make the other negative.

- A sequence that counts down is harder, but a sequence that goes up and then down is a great deal harder. This requires large b and or c and a small negative a.

When constructing these problems, a spreadsheet can be really helpful as it lest you adjust a, b, and c and see the resulting problem immediately.

2.2.1   SOLUTION METHODS

The **first finite difference** of a sequence s_n is defined to be

$$\Delta s_n = s_n - s_{n-1}.$$

The **second finite difference** of a sequence is the finite difference of the first finite differences. For example, given the sequence 1, 3, 7, 13, 21, the first finite differences are

$$3 - 1 = 2, 7 - 3 = 4, 13 - 7 = 6, 21 - 13 = 8$$

and we list them as 2, 4, 6, 8. Then the second finite differences are the finite differences of 2, 4, 6, 8 are

$$4 - 2 = 2, 6 - 4 = 2, 8 - 6 = 2$$

and we list them as 2,2,2.

Here is a fact that can be used to tell if a sequence has a quadratic rule.

The second finite difference of a sequence with quadratic generator rule has **constant value**. If you want to know why, here is a proof.

Proof: Compute the second finite differences of $s_n = a \cdot n^2 + b \cdot n + c$:

$$\begin{aligned} \Delta s_n &= s_n - s_{n-1} \\ &= a \cdot n^2 + b \cdot n + c - a \cdot (n-1)^2 - b(n-1) - c \\ &= 2an - a + b \end{aligned}$$

so

$$\begin{aligned} \Delta^2 \text{ (the second finite difference) } s_n &= \Delta(2an - a + b) \\ &= 2an - a + b - 2a(n-1) + a - b \\ &= 2a. \end{aligned}$$

Notice that a is half of the constant value we find when computing the second difference. Once we know a we can find b. The value of c is simply s_0, which can be found by substituting $n = 0$ into s_n.

Example 2.3   Start with the quadratic-rule sequence 2, 4, 8, 14, 22, ... and compute differences. The first finite difference is $4 - 2$, $8 - 4$, $14 - 8$, $22 - 14$, ... which is 2, 4, 6, 8, ... The second finite difference is then 2, 2, 2, ... This tells us $a = 1$. This means the quadratic part of the sequence is $1 \cdot n^2 = n^2$. Since the first term is 2, that means when $n = 1$ $s_1 = 4 = 1^2 + b \cdot (1) + 2$. From here we can deduce that $b = 1$ telling us that the sequence formula is

$$s_n = n^2 + n + 2.$$

2.3   RECURRENCE RELATION GENERATORS

The next type of sequence is one that uses a simple rule, but often not one that can be turned into a function used to generate the terms of the sequence. One of the sequences that can be generated in this fashion is the famous Fibonacci numbers:

$$1, 1, 2, 3, 5, 8, 13, 21, 34, \ldots.$$

The rule for the Fibonacci numbers is that the next number is the sum of the last two. That rule cannot be applied for the first two numbers, so we need a starting point. For the Fibonacci numbers, we use 1, 1 as the first two numbers, giving us a starting point. If f_n is the nth Fibonacci number then we can state the rule, called the *recurrence relation*, as

$$f_n = f_{n-1} + f_{n-2}.$$

A recurrence relation-based sequence requires a rule for how to compute the next numbers from the numbers that come before it in the sequence. That rule is called the *recurrence relation* for the sequence. The *order* of the recurrence relation is the number of previous terms that the next number depends on. The Fibonacci numbers, for example, are *order 2* or *second order*. Once we have the recurrence relation, then we also need the starting numbers which are called the *initial conditions*, like the 1,1 we use to generate the Fibonacci numbers.

Example 2.4   Suppose that we have a sequence with the recursion

$$g_2 = g_{n-1} + 2g_{n-2}$$

and the initial conditions $g_0 = g_1 = 1$, which is the same as the Fibonacci numbers. We read this as "to generate the third number in the sequence, take the second number and add it to two times the first number." If we wanted the fourth number, we would take the third number and add it to two times the second number, and so on. The sequence is

$$1, 1, 3, 5, 11, 21, 43, \ldots.$$

2.3.1   SOLUTION METHODS

Once we know how to generate the sequence, it is easy to pose sequence extension problems based on the sequence. As long as the recursion relation is based on whole numbers and the initial conditions are whole numbers we get a sequence of whole numbers. There is something interesting under the surface of these sequences. Let's begin by noting that the Fibonacci numbers and the sequence in Example 2.4 have closed formulas:

$$f_n = \frac{1}{\sqrt{5}}\left[\left(\frac{1+\sqrt{5}}{2}\right)^{n+1} - \left(\frac{1-\sqrt{5}}{2}\right)^{n+1}\right]$$

and

$$g_n = \frac{2^n + (-1)^n}{3}.$$

Both of these closed formulas for the sequences can be verified by plugging in $n = 0, 1, 2, 3, \ldots$ to the formulae given. If you are posing this sort of problem, it is possible to design the closed form to some degree. Notice that the Fibonacci numbers have a more complex looking form that Example 2.4 does. Here is the rule for finding the closed form of these sequences.

If a and b are coefficients and

$$h_n = a \cdot h_{n-1} + b \cdot h_{n-2}$$

then the closed form is based on the polynomial $x^2 - ax - b$. If the polynomial has two different roots r_1 and r_2 then

$$h_n = C \cdot r_1^n + D \cdot r_2$$

while if the polynomial has one root r then

$$h_n = C \cdot r^n + D \cdot n \cdot r^n,$$

where C and D are numbers that are calculated from the initial conditions. To find C and D you write the system of linear equations

$$\begin{aligned} h_0 &= C + D \\ h_1 &= C \cdot r_1 + D \cdot r_2 \end{aligned}$$

or

$$\begin{aligned} h_0 &= C \\ h_1 &= C \cdot r + D \cdot r \end{aligned}$$

as appropriate for the values of C and D.

2.4   HAILSTONE-LIKE PROBLEMS

The *hailstone sequence* is part of an unsolved math problem. The main problem is to prove that, no matter what positive whole number you start with, the sequence goes to one. Given a number, n, following terms of the sequence are generated with the following rule:

$$\text{Next term} = \begin{cases} 3n + 1 & n \text{ odd} \\ n/2 & n \text{ even.} \end{cases} \tag{2.1}$$

In other words, if the current number is odd, triple it and add one, but if it is even, divide it in half.

Example 2.5   Compute the first ten terms of the hailstone sequence starting with $n = 29$.

Solution: $29, 88, 44, 22, 11, 34, 17, 52, 26, 13, \ldots$.

It is possible to squeeze many sequence extension problems out of the hailstone sequence by simply starting with different numbers. Notice that the hailstone rule more than triples odd numbers but only divides even numbers in half—and yet mathematicians have verified that for "almost all" numbers it eventually arrives at 1: how is this possible? Look and the 2nd–5th terms of the solution to Example 2.5. Where each odd term is taken immediately to an even term, even terms may lead to even terms. The value 88 is divisible by 2 three times. This means that the tendency of the sequence to grow or shrink is not at all clear. The hailstone sequence sometimes grows to huge values before shrinking, irregularly, down to one.

We can get many more problems by learning to build more rules like the hailstone rule. The trick is to write rules that return a whole number no matter what the input is. A simple way to do this is to use a rule like this:

$$\begin{cases} an + q & n \text{ odd} \\ n/2 & n \text{ even,} \end{cases} \tag{2.2}$$

where a and q are odd numbers. Choosing $a = 3$ and $q = 1$ gives us the original hailstone sequence.

Example 2.6   Starting with 5, generate a sequence using the hailstone-like rule with $a = 3$ and $q = -1$,

$$\begin{cases} 3n - 1 & n \text{ odd} \\ n/2 & n \text{ even} \end{cases} \tag{2.3}$$

$5, 14, 7, 20, 10, 5, 14, 7, 20, 10, \ldots$.

Notice that these numbers form a cycle. As far as we know the original hailstone sequence has only one cycle: $1, 4, 2, 1, 4, 2, \ldots$, the sequence with $a = 3$ and $q = -1$ has at least two cycles in it. The one starting with 5 and also $1, 2, 1, 2, 1, 2, \ldots$.

The issue of cycles in these sequences matters because, if the sequence cycles on a short period the sequence extension problem becomes really easy. You need to avoid short cycles, unless you want to pose an easy problem.

To make one of these rules work, the odd number a must be at least 3 and if it's too large the sequence becomes unmanageable. The number q affords more freedom. Let's look at the example with $a = 5, q = 3$.

Example 2.7   Look at the first ten terms of the hailstone-like sequence, starting with 1, then 3, then 5, using $a = 5$ and $q = 3$:

$$\begin{cases} 5n + 3 & n \text{ odd} \\ n/2 & n \text{ even.} \end{cases} \tag{2.4}$$

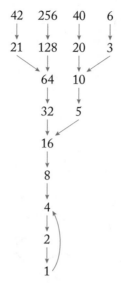

Figure 2.1: The inverse tree of the hailstone sequence.

Starting with 1: 1, 8, 4, 2, 1, 8, 4, 2, 1, 8, ...

Starting with 3: 3, 18, 9, 48, 24, 12, 6, 3, 18, 9, ...

Starting with 5: 5, 28, 14, 7, 36, 18, 9, 48, 24, 12, ...

This rule is rich in cycles.

2.4.1   SOLUTION METHODS

There is almost no general structure to the problems that arise from hailstone-like functions and so the solution method is to simply look for the pattern. There is a construct you can use to help students understand these, the inverse-tree of the rule. This tree starts with a number at the bottom and then grows upward by showing which numbers will generate that number under the influence of the rule. An example, for the hailstone sequence, is shown in Figure 2.1.

Hailstone-like sequences are, if you can manage to code the rule in a spreadsheet, another place where a spreadsheet is a wonderful design aid. In Excel, the **IF** function is key.

2.5   STATE-CONDITIONED SEQUENCES

An example of a finite state controller (FSC) used to generate sequences is shown in Figure 2.2. The arrow pointing to state A gives the initial value of the sequence. After that you follow the arrows and do what they say.

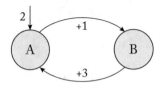

Figure 2.2: An example of a finite state control.

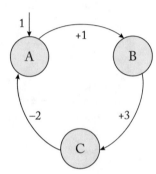

Figure 2.3: An example of a finite state control with three states.

Example 2.8   The FSC given in Figure 2.2 generates the sequence 2, 3, 6, 7, 10, 11, 14, 15, 18, 19, …. The sequence starts with two and then alternates adding one and three to generate the next term. The "2", "+1", and "+3" in the machine in Figure 2.2 can be replaced with other numbers.

If we change the starting number to 5, the A to B transition from +1 to +2 and the B to A transition from +3 to +4 then the machine generates the sequence 5, 7, 11, 13, 17, 19, 23, 25, 29, 31, ….

   This first example of a finite state controller, taking three numerical parameters, is itself a mini-problem factory. Any three numbers filled in generate a sequence and, from it, we get sequence completion problems. This controller has no branch-points in its arrows; adding these can give us much more complex patterns. Before we move onto these more sophisticated controllers, note that the A/B two-state controller can be made more complex, and made to yield more complex patterns, by adding more states.

Example 2.9   The FSC given in Figure 2.3 generates the sequence 1, 2, 5, 3, 4, 7, 5, 6, 9, 7, 8, ….

Adding the −2 in the C to A transition means that the sequence can back up. The hailstone-type sequences are completely determined by their current value—the FSC-based sequences are not. Notice that the sequence above contains two 7's but one moves on to 5 and the other moves on to 8. In fact, since the sum around the ring is $+1 + 3 - 2 = 2$, the numbers will increase

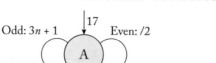

Figure 2.4: An example of a finite state control with a branch.

indefinitely, but not monotonically. The ability to have a repeated value that does not send the sequence into a cycle is because of the memory the sequence has of which state it is in. If we replace $1, +2, +3, -2$ by $3, +2, +2, +1$ we generate the following sequence: 1, 3, 5, 6, 8, 10, 11, 13, 15, 16, …. The pattern, after the initial value of 1 is "add two, add two, add one."

These FSCs, with no branches in their control strategy, can have as many states as you want. This amounts to establishing a simple, periodic pattern of increments. Having decrements (-2) makes the pattern a little stranger. Throwing in the occasional zero can also create at least temporary confusion. Using more than four states can create a pattern that is a little hard to see, so use caution when giving these to students.

If we put branches into the FSC's then we are able to create much more complex patterns. A *branch* in an FSC can be created by any test on the current number. If the test is true, we follow one branch out of the state. It is important to make sure there is a branch for every possibility. The tests for "odd or even" form an example that has a true outcome for all numbers. These tests let us capture the kind of rules that appear in the hailstone-like sequences. A simple machine, with only one state, that mimics the hailstone sequence starting at $n = 17$ appears in Figure 2.4.

Example 2.10 The sequence specified by the machine in Sequence 2.4 is 17, 52, 26, 13, 40, 20, 10, 5, 16, 8, …, it exactly matches the original hailstone sequence. It divides even numbers by two and takes odd numbers, triple them, and add one.

An FSC that has states with branches can generate very complex sequences. With one state, for example, we can simulate the hailstone sequence. An FSC can have both branched and unbranched states, like the one in Figure 2.5.

Example 2.11 The sequence specified by the machine in Sequence 2.5 is 1, 3, 7, 8, 9, 11, 15, 5, 6, 7, 9, 13, 14, 15, 17, 21, 7, 9, 11, 15, 5, 6, 7, 9, 13, …. The pattern is not at all obvious, this is why the example part of the sequence is longer in this example. It is important that there be several examples of the divide-by-three part of the rule.

When an FSC has multiple branched states the patterns it generates can be extremely complex. In the next example we will look at FSC with two states, both of which are branched. The basic pattern for this machine is alternating between two rules for odd numbers, one of which is the classical hailstone rule. The FSC with two branched states appears in Figure 2.6.

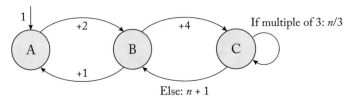

Figure 2.5: An example of a finite state control with a branch. The branch is in State C. If the number is a multiple of 3 when the sequence is in state 3, the next number is calculated by dividing by three, otherwise one is added.

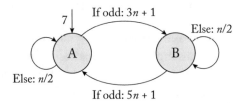

Figure 2.6: An example of a finite state control with two branched states.

Example 2.12   The sequence specified by the machine in Sequence 2.6 is 7, 22, 11, 56, 28, 14, 7, 22, 11, 56, 28, 14, 7, ..., a fairly short cycle.

The machine in Figure 2.6 is similar to the hailstone sequence, but it takes an odd number to $3n + 1$ and $5n + 1$ alternately. The examples we have presented in this section only scratch the surface of the possible branching rules and the functions that can be applied. Because we want to stick with whole number sequences, it is important to make sure that only even numbers are divided by two, only multiples of three are divided by three, and so on. In the examples so far in this section, constraints like divisibility are handled by using rules that explicitly test for the divisibility with the branching rules, e.g., "If the number is a multiple of two, divide by two."

We conclude with an example that uses a greater variety of arithmetic operations, shown in Figure 2.7. Notice that the number arriving in state B is always odd and, as a result, the number arriving in state C is always even. This sort of control is key to planning FSCs.

Example 2.13   The sequence generated by the machine in Figure 2.7 is 1, 4, 7, 12, 25, 11, 44, 47, 52, 105, 51, 204, 207, 212, 425, 211, 844, 847, 852, 1705, ...

This sequence is not to difficult. It is easy to notice the $+3$ then $+5$ arc which is always followed by $2n + 1$ and then $(n - 3)/2$. With this information the only point for confusion is the possibly optional $4n$ in state A.

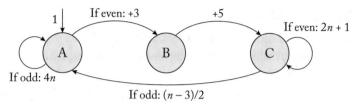

Figure 2.7: A more complex finite state controller with a variety of operations.

2.5.1   SOLUTION METHODS

For the non-branching FSCs the problems can be solved by looking for cyclic patterns in the sequence or, more effectively, at differences. Let's compute the difference of the sequence in Example 2.8. The sequence was $2, 3, 6, 7, 10, 11, 14, 15, 18, 19, \ldots$. If we compute the differences between terms we get $1, 3, 1, 3, 1, 3, 1, 3, 1, \ldots$, which makes what is going on very obvious. The length of the repeating pattern in the difference sequences is equal to the number of states in a non-branching FSC.

For branching FSCs we are in the same situation as the hailstone sequence. No single, simple analysis technique will work and so looking for patterns, e.g., "sometimes things that are a multiple of three are divided by three" and then building on those patterns can help a student figure out the pattern.

As Example 2.13 shows, including a short segment of changes (the $+3$, $+5$) that is easy to spot is a way of creating a place where one of these FSC-controlled sequences begins to unravel. In general, FSC-based sequence extension problems require more examples than the earlier sorts of problems. When using FSC sequence generators that have branching it is very important to run the sequence forward yourself to make sure you have not created a monster. Some monsters are just impossible to figure out, others simply grow so fast that they rapidly get out of control. You want to avoid overflowing the student's calculators.

2.6   SOME OTHER GENERATING METHODS, SEQUENCES, AND PROBLEMS

So far in this chapter we have seen five different methods of generating sequence completion problems with varying degrees of difficulty. This section gives some simple techniques for generating more sequence completion problems from those you already have.

Possibly the simplest way of varying a sequence is to run the whole sequence through a linear transformation, starting with $S = \{s_1, s_2, s_3, \ldots\}$ and generating a new sequence T with $t_i = a \cdot s_i + b$ for whole numbers a, b.

2.6.1   SHUFFLE THE DECK

A very simple method of generating a much harder sequence is to take every other entry from two different sequences, at least usually. Occasionally shuffling two sequences together will yield a sequence that follows a single, simple rule, but this is rare. Typically, a larger number of sequence values—about twice as many—should be given for shuffled sequences. You can, of course, shuffle as many sequences together as you want, but shuffling more than two together really thins out the available information to where the problem is likely to be excessively hard.

Something to consider when shuffling sequences together is that making one of the sequences more or less trivial is a way of making an easier shuffled sequence problem. Giving the sequence 1, 1, 1, 1, 1, ... or even 1, 2, 3, 4, 5 , ... is probably too easy, but using one of these as half of a shuffle simply turns up the difficulty of the other shuffled sequence a little.

It is also possible to make irregular shuffles. Look at the binary sequence 0, 0, 1, 0, 0, 1, 0, 0, 1, ... If you use a number from one sequence to replace the 0s and a number from another sequence to replace the 1s that gives you an irregular shuffle. A standard shuffle is based on the sequence 0, 1, 0, 1, 0, 1, 0, 1, ...

2.6.2   A FEW OTHER SEQUENCES

There are sequences that arise naturally in mathematics or which arise from stupid tricks. A natural mathematical sequence is the prime numbers, 2, 3, 5, 7, 11, 13, 17, 19, 23, 29, ...

Another natural sequence is the "spoken sequence." This sequence starts 1 and then you read it "one one," "two one," "one two one one," "one one one two two one," "three one two two one one," and so on. Numerically this sequence is 1, 11, 21, 1211, 111221, 312211, 13112211, 1113212221, 311312113211, ... This sequence is linguistic rather than mathematical which can make it quite difficult for mathematically sophisticated students. A spoken sequence can be started with any digit or collection of digits. If we start with 2 we get 2, 12, 1112, 3112, 132112, 1113122112, 311311222112, ...

Negative membership sequences are also potentially challenging. These are sequences that *do not* contain certain things. An example of this sort of sequence is to take the natural numbers but exclude all multiples of 2 or 3, which gives us the sequence 1, 5, 7, 11, 13, 17, 19, 23, 25, 29, ... A more complex negative membership sequence is to take the natural numbers but exclude any number that has a perfect square as a divisor. This would give us 1, 2, 3, 5, 6, 7, 10, 11, 13, 14, 15, 17, 19, 21, 22, 23, ...

If a sequence is increasing, that is to say if the next entry is always larger than the last one, then you can use that sequence as the exclusion rule for a negative membership sequence. Suppose we take the sequence with the rule $s_i = 3 \cdot i + 1$ for $i \geq 1$. This given use the sequence 4, 7, 10, 13, 16, 19, 22, 25, 28, 31, ... Using this as an excluded set, we count up from one skipping the members of this sequence. This yields 1, 2, 3, 5, 6, 8, 9, 11, 12, 14, ... Not too difficult except for a slightly irregular start, which may throw the students off the scent briefly. The sequence consisting of everything that is not a Fibonacci number would be harder.

2.6.3 OTHER PROBLEMS ON SEQUENCES

A way to make the problem different is to give a sequence with one or more entries before the end missing and ask the students to fill in the missing entries. Another natural variation is to ask the student to find the rule for the entire sequence. This can be anywhere from easy, with linear sequences, to quite difficult for some of the finite state controller sequences. Keep in mind that there may be more than one reasonable way to state the rule for a sequence and do not just check student's answers against our keys or the rule you devised to generate the sequence. If you modify a sequence problem, do double check that it is still a reasonable, solvable problem.

2.7 EXAMPLE PROBLEMS

This section concludes Chapter 2 with a page each of examples of each of the five types of sequence generators we have covered. This is a source of quick problems that we have already tested, in case you need problems in a hurry. Some of the problem types give the solution, others, where it is not hard to compute the solution, do not, in part to make room for a few more problems. We did our best to check that the sequences and solutions, when given, are correct. That said, it is never a bad idea to check problems you assign for correctness. Also, if you find an error, we would love to hear about it care of our publishers, Morgan & Claypool.

Simple Linear Rules

Sequence: 6,9,12,15,18,...
Solution: 21
Rule: 3×n+3

Sequence: 3,5,7,9,11,...
Solution: 13
Rule: 2×n+1

Sequence: 11,17,23,29,35,...
Solution: 41
Rule: 6×n+5

Sequence: 10,19,28,37,46,...
Solution: 55
Rule: 9×n+1

Sequence: 6,14,22,30,38,...
Solution: 46
Rule: 8×n-2

Sequence: 1,8,15,22,29,...
Solution: 36
Rule: 7×n-6

Sequence: -3,-2,-1,0,1,...
Solution: 2
Rule: n-4

Sequence: 4,11,18,25,32,...
Solution: 39
Rule: 7×n-3

Sequence: 1,5,9,13,17,...
Solution: 21
Rule: 4×n-3

Sequence: 20,35,50,65,80,...
Solution: 95
Rule: 15×n+5

Sequence: 23,41,59,77,95,...
Solution: 113
Rule: 18×n+5

Sequence: 13,21,29,37,45,...
Solution: 53
Rule: 8×n+5

Sequence: 24,32,40,48,56,...
Solution: 64
Rule: 8×n+16

Sequence: 24,43,62,81,100,...
Solution: 119
Rule: 19×n+5

Sequence: 55,115,175,235,295,...
Solution: 355
Rule: 60×n-5

Sequence: 38,83,128,173,218,...
Solution: 263
Rule: 45×n-7

Sequence: 60,126,192,258,324,...
Solution: 390
Rule: 66×n-6

Quadratic Rules
Sequence: 6,18,36,60,90,...
Solution: 126
Rule: $3 \times n^2 + 3 \times n + 0$

Sequence: 8,22,42,68,100,...
Solution: 138
Rule: $3 \times n^2 + 5 \times n$

Sequence: 9,12,17,24,33,...
Solution: 44
Rule: $n^2 + 8$

Sequence: 2,5,10,17,26,...
Solution: 37
Rule: $n^2 + 1$

Sequence: 6,14,24,36,50,...
Solution: 66
Rule: $n^2 + 5 \times n$

Sequence: 6,14,26,42,62,...
Solution: 86
Rule: $2 \times n^2 + 2 \times n + 2$

Sequence: 11,21,37,59,87,...
Solution: 121
Rule: $3 \times n^2 n + 7$

Sequence: 12,20,30,42,56,...
Solution: 72
Rule: $n^2 + 5 \times n + 6$

Sequence: 7,11,17,25,35,...
Solution: 47
Rule: $n^2 + n + 5$

Sequence: 3,7,13,21,31,...
Solution: 43
Rule: $n^2 + n + 1$

Sequence: 6,21,42,69,102,...
Solution: 141
Rule: $3 \times n^2 + 6 \times n - 3$

Sequence: 2,9,22,41,66,...
Solution: 97
Rule: $3 \times n^2 - 2 \times n + 1$

Sequence: 2,13,28,47,70,...
Solution: 97
Rule: $2 \times n^2 + 5 \times n - 5$

Sequence: 3,15,33,57,87,...
Solution: 123
Rule: $3 \times n^2 + 3 \times n - 3$

Sequence: 9,15,27,45,69,...
Solution: 99
Rule: $3 \times n^2 - 3 \times n + 9$

Sequence: 59,48,35,20,3,...
Solution: -16
Rule: $n^2 - 8 \times n + 68$

Sequence: 60,54,50,48,48,...
Solution: 50
Rule: $n^2 - 9 \times n + 68$

Simple Recurrence Relations

Sequence: 3,3,15,39,123,363,...
Rule: $s_n = 2s_{n-1} + 3s_{n-2}$
Starts with: 3 3

Sequence: 2,1,3,4,7,11,...
Rule: $s_n = s_{n-1} + s_{n-2}$
Starts with: 2 1

Sequence: 4,2,14,46,166,590,...
Rule: $s_n = 3s_{n-1} + 2s_{n-2}$
Starts with: 4 2

Sequence: 1,3,15,69,321,1491,...
Rule: $s_n = 4s_{n-1} + 3s_{n-2}$
Starts with: 1 3

Sequence: 2,4,16,48,160,512,...
Rule: $s_n = 2s_{n-1} + 4s_{n-2}$
Starts with: 2 4

Sequence: 3,3,21,75,309,1227,...
Rule: $s_n = 3s_{n-1} + 4s_{n-2}$
Starts with: 3 3

Sequence: 3,1,15,49,207,817,...
Rule: $s_n = 3s_{n-1} + 4s_{n-2}$
Starts with: 3 1

Sequence: 1,2,11,50,233,1082,...
Rule: $s_n = 4s_{n-1} + 3s_{n-2}$
Starts with: 1 2

Sequence: 2,4,10,24,58,140,...
Rule: $s_n = 2s_{n-1} + s_{n-2}$
Starts with: 2 4

Sequence: 3,2,3,7,18,47,...
Rule: $s_n = 3s_{n-1} - s_{n-2}$
Starts with: 3 2

Sequence: 1,1,6,21,81,306,...
Rule: $s_n = 3s_{n-1} + 3s_{n-2}$
Starts with: 1 1

Sequence: 4,3,5,12,31,81,...
Rule: $s_n = 3s_{n-1} - s_{n-2}$
Starts with: 4 3

Sequence: 2,1,7,10,31,61,...
Rule: $s_n = s_{n-1} + 3s_{n-2}$
Starts with: 2 1

Sequence: 4,4,12,20,44,84,...
Rule: $s_n = s_{n-1} + 2s_{n-2}$
Starts with: 4 4

Sequence: 3,2,-1,-3,-2,1,...
Rule: $s_n = s_{n-1} - s_{n-2}$
Starts with: 3 2

Sequence: 2,1,7,23,83,295,...
Rule: $s_n = 3s_{n-1} + 2s_{n-2}$
Starts with: 2 1

Sequence: 1,2,6,16,44,120,...
Rule: $s_n = 2s_{n-1} + 2s_{n-2}$
Starts with: 1 2

Hailstone-Like Sequences

The starting point for a hailstone-like sequence is simply its first number, all that is needed to pose the problem is the sequence itself; the rule gives the solution.

Sequence: 7, 22, 11, 34, 17, 52, 26, 13, 40...

Rule: $\begin{cases} 3n + 1 & n \text{ odd} \\ n/2 & n \text{ even} \end{cases}$

Sequence: 9, 26, 13, 38, 19, 57, 56, 28, 14...

Rule: $\begin{cases} 3n - 1 & n \text{ odd} \\ n/2 & n \text{ even} \end{cases}$

Sequence: 15, 78, 39, 198, 99, 498, 249, 1248, 624, 312 ...

Rule: $\begin{cases} 5n + 3 & n \text{ odd} \\ n/2 & n \text{ even} \end{cases}$

Seq: 7, 36, 18, 9, 46, 23, 116, 58, 29, 146, ...

Rule: $\begin{cases} 5n + 1 & n \text{ odd} \\ n/2 & n \text{ even} \end{cases}$

Seq: 3, 14, 7, 26, 13, 44, 22, 11, 38, 19, ...

Rule: $\begin{cases} 3n + 5 & n \text{ odd} \\ n/2 & n \text{ even} \end{cases}$

Seq: 29, 88, 44, 22, 11, 34, 17, 52, 26, 13,...

Rule: $\begin{cases} 3n + 1 & n \text{ odd} \\ n/2 & n \text{ even} \end{cases}$

Sequence: 15, 44, 22, 11, 32, 16, 8, 4, 2, 1...

Rule: $\begin{cases} 3n - 1 & n \text{ odd} \\ n/2 & n \text{ even} \end{cases}$

Seq: 17, 88, 44, 22, 11, 58, 29, 148, 74, 37, ...

Rule: $\begin{cases} 5n + 3 & n \text{ odd} \\ n/2 & n \text{ even} \end{cases}$

Seq: 11, 56, 28, 14, 7, 36, 18, 9, 46, 23, ...

Rule: $\begin{cases} 5n + 1 & n \text{ odd} \\ n/2 & n \text{ even} \end{cases}$

Seq: 15, 50, 25, 80, 40, 20, 10, 5, 20, 10, ...

Rule: $\begin{cases} 3n + 5 & n \text{ odd} \\ n/2 & n \text{ even} \end{cases}$

Seq: 3, 18, 9, 60, 30, 15, 102, 51, 354, 177, ...

Rule: $\begin{cases} 7n - 3 & n \text{ odd} \\ n/2 & n \text{ even} \end{cases}$

Sequence: 5, 18, 9, 30, 15, 48, 24, 12, 6, 3, ...

Rule: $\begin{cases} 3n + 3 & n \text{ odd} \\ n/2 & n \text{ even} \end{cases}$

Seq: 3, 14, 7, 34, 17, 84, 42, 21, 104, 52, ...

Rule: $\begin{cases} 5n - 1 & n \text{ odd} \\ n/2 & n \text{ even} \end{cases}$

Sequence: 102, 51, 252, 126, 63, 312, 156, 78, 38, 19, ...

Rule: $\begin{cases} 5n - 3 & n \text{ odd} \\ n/2 & n \text{ even} \end{cases}$

Sequence: 1, 10, 5, 12, 6, 3, 16, 8, 4, 2, ...

Rule: $\begin{cases} 3n + 7 & n \text{ odd} \\ n/2 & n \text{ even} \end{cases}$

Sequence: 7, 46, 23, 158, 79, 550, 275, 1922, 961, 6724, ...

Rule: $\begin{cases} 7n - 3 & n \text{ odd} \\ n/2 & n \text{ even} \end{cases}$

Seq: 44, 22, 11, 36, 18, 9, 30, 15, 48, 24, ...

Rule: $\begin{cases} 3n + 3 & n \text{ odd} \\ n/2 & n \text{ even} \end{cases}$

Seq: 9, 44, 22, 11, 54, 27, 134, 67, 334, 167, ...

Rule: $\begin{cases} 5n - 1 & n \text{ odd} \\ n/2 & n \text{ even} \end{cases}$

Sequence: 11, 52, 26, 13, 62, 31, 152, 76, 38, 19, ...

Rule: $\begin{cases} 5n - 3 & n \text{ odd} \\ n/2 & n \text{ even} \end{cases}$

Seq: 9, 24, 17, 58, 29, 94, 47, 148, 74, 37, ...

Rule: $\begin{cases} 3n + 7 & n \text{ odd} \\ n/2 & n \text{ even} \end{cases}$

Seq: 19, 52, 26, 13, 34, 17, 46, 23, 64, 32, ...

Rule: $\begin{cases} 3n - 5 & n \text{ odd} \\ n/2 & n \text{ even} \end{cases}$

FSC-Generated Sequences

Seq: 3, 5, 10, 12, 17, 19, 24, 26, 31, 33, . . .
Machine: Figure 2.2 with 3, +2, +5.
Solution: 38.

Seq: 1, 4, 8, 11, 15, 18, 22, 25, 29, 32, . . .
Machine: Figure 2.2 with 1, +3, +4.
Solution: 36.

Seq: 6, 9, 8, 11, 10, 13, 12, 15, 14, 17, . . .
Machine: Figure 2.2 with 6, +3, -1.
Solution: 16

Seq: 1, 7, 14, 20, 27, 33, 40, 46, 53, 59, . . .
Machine: Figure 2.2 with 1, +6, +7.
Solution: 66

Seq: 5, 3, 8, 6, 11, 9, 14, 12, 17, 15, . . .
Machine: Figure 2.2 with 5, -2, +5.
Solution: 20

Seq: 1, 3, 4, 6, 8, 9, 11, 13, 14, 16, . . .
Machine: Figure 2.3 with 1, +2, +1, +2.
Solution: 18

Seq: 2, 3, 5, 6, 7, 9, 10, 11, 13, 14, . . .
Machine: Figure 2.3 with 2, +1, +2, +1.
Solution: 15

Seq: 5, 4, 7, 9, 8, 11, 13, 12, 15, 17, . . .
Machine: Figure 2.3 with 5, -1, +3, +2.
Solution: 16

Seq: 4, 6, 9, 13, 15, 18, 22, 24, 27, 31, . . .
Machine: Figure 2.3 with 4, +2, +3, +4.
Solution: 33

Seq: 1, 3, 8, 7, 9, 14, 13, 15, 20, 19, . . .
Machine: Figure 2.3 with 1, +2, +5, -1.
Solution: 21

Seq: 2, 4, 8, 9, 10, 12, 16, 17, 18, 20, 24,
8, 9, 10, 12, 16, 17, 18, 20, . . .
Machine: Figure 2.5 starting with 2
Solution: 24

Seq: 8, 10, 14, 15, 16, 18, 22, 23, 24, 26, 30,
10, 11, 12, 14, 18, 6, 2, 3, 4, . . .
Machine: Figure 2.5 starting with 8.
Solution: 6

Seq: 11, 13, 17, 18, 19, 21, 25, 26, 27, 29,
33, 11, 12, 13, 15, 19, 20, 21, 23, 27, . . .
Machine: Figure 2.5 starting with 11.
Solution: 9

Seq: 11, 34, 17, 86, 43, 130, 65, 326, 163, 490,
245, 1226, 613, 1840, 920, 460, 230, 115,..
Machine: Figure 2.6 starting with 11.
Solution: 576

Seq: 76, 38, 19, 58, 29, 146, 73, 220, 110, 55,
276, 138, 69, 208, 104, 52, 26, 13, 66, 33,..
Machine: Figure 2.6 starting with 76.
Solution: 100

Seq: 23, 70, 35, 176, 88, 44, 22, 11, 34, 17, 86,
43, 130, 65, 326, 163, 490, 245, 1226,
613, . . .
Machine: Figure 2.6 starting with 23.
Solution: 1840

Seq: 2, 5, 10, 21, 9, 36, 39, 44, 89, 43, 172,
175, 180, 361, 179, 716, 719, 724, 1449,
723, . . .
Machine: Figure 2.7 starting with 2.
Solution: 2892

Seq: 3, 12, 15, 20, 41, 19, 76, 79, 84, 169, 83,
332, 335, 340, 681, 339, 1356, 1359, 1364,
2729, . . .
Machine: Figure 2.7 starting with 3.
Solution: 1363

Seq: 4, 7, 12, 25, 11, 44, 47, 52, 105, 51, 204,
207, 212, 425, 211, 844, 847, 852, 1705,
851, . . .
Machine: Figure 2.7 starting with 4.
Solution: 3404

CHAPTER 3

Basic Analytic Geometry Problems

Analytic geometry is the application of algebra, calculus, or analysis to problems in geometry. In this chapter we will stick pretty close to the algebra end of the discipline. Many of the problems in this section are not only problem factories, they also work as examples for introductory algebra or geometry classes. They might also prove useful as review problems.

3.1   PROBLEMS ABOUT RIGHT ANGLES

We will start with the fact that the slopes of lines are perpendicular are negative reciprocals of one another. The skills used are finding the slope of a line between a pair of points as well as problem solving and deduction.

Example 3.1   Are the points (1,4), (3,1), (6,3) the vertices of a right triangle?

Solution: Examine the triangle in Figure 3.1.

If there is a right angle, it is the lowest of the three angles. Compute the two slopes:

$$\frac{1-4}{3-1} = -\frac{3}{2}$$

$$\frac{3-1}{6-3} = \frac{2}{3}.$$

Since the slopes are negative reciprocals, the bottom angle is a right angle and the points are the vertices of a right triangle.

The obvious problem factory here is placing three points and asking if they are the vertices of a right triangle—and the answer can be yes or no. This kind of problem can be made much larger. Several instances of this sort of problem can be posed together.

Example 3.2   How many right triangles appear with vertices that are labeled points in Figure 3.2?

Answer: 3. The right triangles are (0,0)-(0,5)-(5,5); (0,0)-(5,5)-(6,2); (0,5)-(1,3)-(5,5).

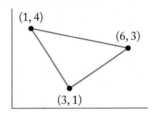

Figure 3.1: A triangle with three marked points.

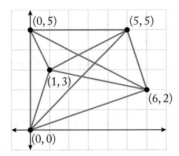

Figure 3.2: This picture shows five points with several right triangles in them.

Notice that drawing the points permits some pairs of lines to be dismissed for narrow angles. Drawing the points and triangle of the motivating question lets you decide which pair of slopes to compute first, at some savings of labor. This is a place where estimation can be shown to save work. In the multi-triangle example the only non-right triangle that is close enough not to be dismissed by eye is (1,3)-(5,5)-(6,2).

3.1.1   SOLUTION TECHNIQUES

When solving the problem that asks if three points are the vertices of a right triangle, simply compute the three slopes and look for negative reciprocal pairs. This is also a time when a calculator may be inferior to just simplifying fractions. Knowing that 0.71428571 and -1.4 are negative reciprocals of one another is harder than knowing the same fact for $\frac{5}{7}$ and $-\frac{7}{5}$. or the problems that ask the number of right triangles that can be made with five points, there is a handy diagram for racking up all ten possible slopes (see Figure 3.3).

Another sort of problem you can do is to ask *are these points the vertices of a square?* This is a little trickier because a square has two sets of conditions; the sides all have to be the same length, but the angles have to be right angles as well! Figure 3.4 shows two quadrilaterals that obey the rule that all sides must be the same length but only the first is a square.

Generating a square is easy. Pick a corner point (c, d) and two positive values a and b. The corners of the square are (c, d), $(c + a, d + b)$, $(c + a - b, d + b + a)$, and $(c - b, d + a)$. The

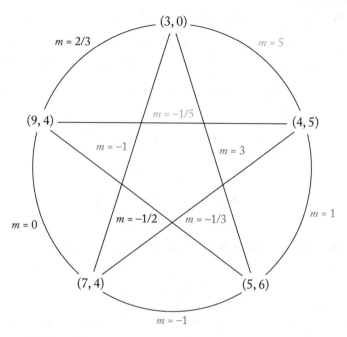

Figure 3.3: A method of diagramming the ten slopes that arise from five points. Note that we have four right angles: two pairs of −1, 1, a pair −1/3, 3, and a pair −1/5, 5. The negative-reciprocal pairs have been given the same color.

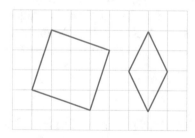

Figure 3.4: Two quadrilaterals with all sides equal, only one of which is a square.

corners of a diamond (make sure $a \neq b$) are (c, d), $(c + a, d + b)$, $(c + 2a, d)$, $(c + a, d − b)$. If $a = b$, then the diamond formula is also a square, which may be useful, but fails to be a square.

For all the problems in this section there is the option of presenting the problem as a list of points or presenting the problem in diagrammatic form. Presenting the problem with a diagram usually reduces the level of difficulty. The directions "make a picture" can bridge this gap. Another point that makes generating problems easy is that scaling, rotation, reflection, and translation all preserve answers to this sort of problem.

3.2   FINDING QUADRATICS THROUGH THREE POINTS

This section deals with a classical analytic geometry problem—given three points, find the quadratic equation whose graph goes through them. Let's start by working an example.

Example 3.3   Find the quadratic equation $y = ax^2 + bx + c$ whose graph goes through the points (1,3), (2,5), and (3,9).

Solution: Plug in the three points (x, y) to the equation and we get the following system of equations:

$$a + b + c = 3 \text{ from } (1, 3)$$
$$4a + 2b + c = 5 \text{ from } (2, 5)$$
$$9a + 3b + c = 9 \text{ from } (3, 9).$$

Subtracting the first equation from the second and the second equation from the third we get:

$$3a + b = 2$$
$$5a + b = 4$$

which reduces by subtracting the first equation from the second to $2a = 2$ or $a = 1$. Plugging the value of a in yields $b = -1$ and plugging both values in yields $c = 3$ so we see the graph of the quadratic function $y = x^2 - x + 3$ contains all three points.

This type of problem builds skills with solving simultaneous equations and permits the students a visual check on their results by graphing the equation with the points, as in Figure 3.5. This is a visceral form of verification that also employs the student's skills at graphing equations.

The choice of x coordinates of 1, 2, and 3 in Example 3.3 was deliberate. If one of the coordinates is $x = 0$ then the problem gets noticeably easier. Let's do another example.

Example 3.4   Find the quadratic equation $y = ax^2 + bx + c$ whose graph goes through the points (0,2), (1,3), and (2,8).

Solution: Plug in the three points (x, y) to the equation and we get the following system of equations:

$$c = 2$$
$$a + b + c = 3$$
$$4a + 2b + c = 8.$$

Having one of the points with x-coordinate zero gives us the value of c immediately: $c = 2$. This gets us to two equations pretty directly.

$$a + b = 1$$
$$4a + 2b = 6.$$

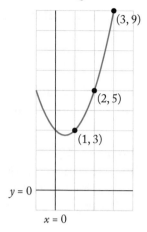

Figure 3.5: The curve $y = x^2 - x + 3$ and the three points used to find it.

Subtracting twice the first equation from the second gives us $2a = 4$ or $a = 2$ which, in turn, gives us $b = -1$ and we see the quadratic equation is $y = 2x^2 - x + 2$.

3.2.1   WHY IS THIS A PROBLEM FACTORY?

Finding the quadratic equation whose graph intersects three points is clearly an analytic geometry problem, but we have not yet made the case that it is a problem factory. For any three points, so long as no two are vertically aligned and all three are not on a horizontal line, there is a unique quadratic equation $y = ax^2 + bx + c$ whose graph intersects the three points. Given this, let's make the factory.

Almost every collection of three points that has a quadratic function passing through those points has coefficients that are not whole numbers. To find a problem where we get three points with whole number coordinates that yield a quadratic equation with whole number coefficients we simply work backward. Pick your answer and plug in three small whole numbers.

So, for example, if we choose $y = x^2 + 2x + 2$ and plug in three values $x = 1$, $x = 2$, and $x = 3$ we get the points (1,5), (2,10), (3,17). Working through the linear equations from those three points:

$$a + b + c = 5 \text{ from } (1, 5)$$
$$4a + 2b + c = 10 \text{ from } (2, 10)$$
$$9a + 3b + c = 17 \text{ from } (3, 17)$$

we will get $a = 1$, $b = 2$, and $c = 2$.

The choice of the x-values plugged into the quadratic give you a great deal of latitude to make the numbers in the simultaneous equations larger or smaller and so the problem harder or easier. Likewise, smaller coefficients in the quadratic used to generate the points yield smaller values and easier problems.

3.2.2   SOLUTION TECHNIQUES

We noted that if one of the three points has an x-coordinate of zero yields an easier problem. The nice thing is that it is always possible to find a problem with one x-coordinate zero by shifting the equation. Here is the procedure.

1. Pick one of the three x values from one of your three points, and it really doesn't matter which one you choose. Then subtract that value from each of the x values. So, for example, if your x values were 1, 2, and 3, you choose 2 and subtract it from each x value, getting $-1, 0$, and 1. This method guarantees you will get one zero in your list of x values.

2. Solve the modified problem, obtaining $g(x) = a'x^2 + b'x + c'$. Then the answer to the unmodified problem, $f(x) = ax^2 + bx + c$ is given by $f(x) = g(x - p)$, where p is the value of x you chose in step 1. This shifts the function back over to its original position.

Example 3.5   Let's demonstrate the technique on the points from Example 3.3: (1,3), (2,5), and (3,9). Choose $p = 2$ and we get the points $(-1, 3)$, $(0,5)$, and $(1,9)$. Plugging these into an abstract quadratic we get the system of equations:

$$a - b + c = 3 \text{ from } (-1, 3)$$
$$c = 5 \text{ from } (0, 5)$$
$$a + b + c = 9 \text{ from } (1, 9).$$

The point where we plugged in $x = 0$ point gives us $c = 5$. This gives us the equations

$$a - b = -2$$
$$a + b = 4.$$

Adding these equations we get $2a = 2$ and so $a = 1$ and thus $b = 3$. This tells us $g(x) = x^2 + 3x + 5$. We now shift back and get $f(x) = g(x - 2) = (x - 2)^2 + 3(x - 2) + 5 = x^2 - x + 3$, which is the answer we got before.

This technique does not necessarily make a problem easier overall, but it does make the simultaneous equations easier to solve. Usually computing $f(x) = g(x - p)$ is not difficult and so the problem gets longer with easier individual steps.

3.3   FINDING SQUARES WITH A GIVEN AREA

The squares we are working with for this problem factory have corners that are points with both coordinates whole numbers. The basis of the problem factory is the question "find a square with area A" which has an answer that subverts the students expectation.

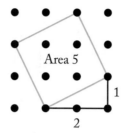

Figure 3.6: A tilted square with area 5 solution.

Since the sides squares can be thought of as built on the vector $\vec{v} = (a, b)$, added in four different orientations to the starting point (c, d), the sides of the square have length $\sqrt{a^2 + b^2}$ and so the square has an area of $a^2 + b^2$. As noted in the Introduction, students are likely to think of squares as aligning with the coordinate axes. If the area of a square is not a perfect (integer) square, then it cannot be aligned with the coordinate axes and have corners with whole number coordinates. The question "find a square of area A," when A is not a perfect integer square, requires a tilted square.

Making "find a square" problems that force tilted squares is actually quite simple. The square can start anywhere and must be based on a vector $\vec{v} = (a, b)$ so that $a^2 + b^2$ is, itself, not a perfect square. Figure 3.6 shows a more detailed diagram of the square of area 5 solution from the introduction. This example, the requirement that it be tilted, follows from $2^2 + 1^2 = 5$, and 5 is not a perfect square. Usually, the interesting problem is finding *Pythagorean triples a, b, c* for which $a^2 + b^2 = c^2$. Perversely, constructing these problems requires that we avoid Pythagorean triples.

A natural next step is to build rectangles that can only be made in tilted form, but this is impossible. A rectangle is defined by its side lengths, $A = a \times b$, so a rectangle of area a can always be drawn in alignment with the coordinate axes. Any factorization of A, including $A \times 1$ permits an aligned solution.

3.4   AREAS ON A GRID

A shaded area on a grid, which is understood to be made of 1×1 squares, can be carefully chosen so that elementary geometric properties can be used to compute the area. More complex areas provide greater geometric challenge.

Motivating question: Assuming the grid shown has a spacing of one, find the shaded area.

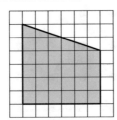

Solution: Divide the shape into a 6 × 4 rectangle and a triangle of width 6 and height 2. The total area is

$$6 \times 4 + \frac{1}{2} \times 6 \times 2 = 30 \text{ units}^2.$$

A shaded-in shape on a grid can yield problems that test a wide variety of skills. The most basic is the rule for the area of a rectangle, closely followed by the area of a triangle. The area formula for triangles—or matching pairs of triangles to make rectangles—adds to the bits of knowledge that these problems can test and reinforce. The area formula for a circle or an arc of a circle can be added to make even more problems.

Question with a semi-circle: Find the shaded area in the diagram below.

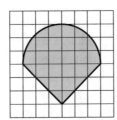

Solution: The half-circle has radius 3 and the triangle has a base of length 6 and a height of three. This means that $A = \frac{1}{2} \cdot \frac{1}{2} \cdot \pi \cdot 3^2 + \frac{1}{2} \cdot 6 \cdot 3 = \frac{9\pi}{4} + 9 = 9\left(\frac{\pi}{4} + 1\right) \cong 16.07$.

It is possible to kick these problems up a notch by requiring the use of logic to resolve overlaps.

Question with overlap: Find the shaded area in the diagram below.

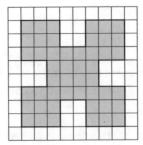

Solution: This shape is made of four 3×3 squares and one 4×4 square. The problem is that each of the smaller squares overlaps the large, central one in a 1×1 region, making the area $4 \cdot 4 + 4 \cdot (3 \cdot 3) - 4 \cdot (1 \cdot 1) = 48$. Since none of the lines in this problem curve or slant, it is also possible to simply *count* the squares. It is important to make the case to the students that the clever solution—adding squares and then subtracting the overlapping areas—is less work. It might also be interesting to note that the shape is an 8×8 square with four 2×2 squares punched out of it: $64 - 4 \times 4 = 64 - 16 = 48$, same answer.

The next problem shows how negative space can be important in solving these problems. While this is a triangle, its base and height are not obvious.

Question solvable with negative space: Find the shaded area in the diagram below.

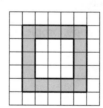

Solution: The shaded area is all of a 5×5 square *except* for three triangles with height and base 3,5; 2,4; and 5,1, respectively yielding area $5^2 - \frac{1}{2} \cdot 3 \cdot 5 - \frac{1}{2} \cdot 2 \cdot 4 - \frac{1}{2} \cdot 5 \cdot 1 = 25 - 14 = 11$.

An obvious technique is to have problems that depend on the difference between two shapes.

Problems with differences of areas: Find the shaded area in the diagram below.

Solution: A 5×5 square less a 3×3 square, $A = 25 - 9 = 16$.

The five example problems show some of the range of the problems that can be posed by shading areas on grids. The additional problems on grid areas at the end of the chapter show many of the other possibilities. As the example with five overlapping squares demonstrates, it is sometimes possible to have a brute force solution. One approach is to present the shaded area with the squares numbered and then decry the effort needed to number all the squares. The idea that knowing the area formulas for rectangles, triangles, and circles or parts of circles leads you to fast, labor-saving calculations of shaded areas.

These problems can simply be posed as fast calculation problems, possibly as a race or other competition. Another way to present the problems is with the directions to find the smallest number of steps that gives the correct area. This is another sort of race, one that takes place in a very different space. This second sort of race is closer to what is done in the formal discipline of mathematics and is preparation for learning the discipline of formal proof.

There is also another, different way to use these problems. You can display two of the problems and ask which shape has the larger area. This changes the character of the problem from exact computation to estimation. A harder version of this is to give several problems and ask students to sort them into ascending or descending order. Exact calculations can be used to do this variation on the areas-on-a-grid problems, but good estimation skills will yield faster results.

3.4.1   USEFUL RULES

Here are the geometry rules used to set up the solutions of all the problem in this section and the example problems at the end of the chapter.

1. The area of a rectangle is width times height.

2. The area of a triangle is one-half base times height.

3. The area of a circle is π times the square of its radius.

4. The area of an arc of a circle is the fraction of 360° full circle subtended by the arc times the full area. A 90° arc, for example, is a quarter of the circle.

3.5   EXAMPLE PROBLEMS

This final section of the chapter gives example problems from each of the sections.

Right Angle Problems

Here are ten sets of points that are the vertices of a right triangle.

1. (1,0), (6,5), (8,3)

2. (2,6), (4,8), (5,3)

3. (0,8), (1,2), (7,3)

4. (1,3), (2,0), (3,2)

5. (1,8), (3,4), (4,7)

6. (4,4), (7,7), (9,5)

7. (3,5), (6,6), (8,0)

8. (0,0), (3,4), (7,1)

9. (3,2), (4,5), (7,4)

10. (0,5), (1,3), (6,8)

Here are ten sets of points that are not the vertices of a right triangle.

1. (2,4), (3,0), (6,2)

2. (2,3), (5,8), (9,3)

3. (3,1), (9,2), (9,3)

4. (0,4), (3,3), (5,3)

5. (4,3), (8,3), (8,4)

6. (0,6), (5,0), (6,6)

7. (0,9), (7,0), (7,7)

8. (2,6), (4,0), (5,3)

9. (2,3), (2,6), (6,4)

10. (3,6), (3,8), (5,5)

Sets of five points with at least two right angles (R.A.).

1. (0 4), (2 9), (4 4), (5 6), (7 5), 2 R.A.

2. (0 5), (2 0), (4 2), (5 8), (7 6), 2 R.A.

3. (1 1), (2 8), (4 1), (5 4), (8 6), 2 R.A.

4. (0 9), (1 2), (3 8), (7 4), (8 3), 2 R.A.

5. (0 2), (1 2), (6 6), (7 6), (8 3), 2 R.A.

6. (0 8), (3 3), (4 2), (6 5), (8 6), 3 R.A.

7. (1 4), (2 1), (5 8), (8 5), (9 4), 3 R.A.

8. (0 0), (1 9), (4 3), (5 4), (6 3), 3 R.A.

9. (0 3), (2 8), (4 0), (5 1), (7 3), 4 R.A.

10. (3 0), (4 5), (5 6), (7 4), (9 4), 4 R.A.

Here are several squares and diamonds.

1. (2,4), (4,1), (5,6), (7,3); square.

2. (2,6), (5,2), (6,9), (9,5); square.

3. (5,6), (6,1), (10,7), (11,2); square.

4. (1,7), (4,4), (4, 10), (7,7); both.

5. (0,9), (2,2), (7,11), (9,4); square.

6. (3,4), (5,0), (5,8), (7,4); diamond.

7. (2,2), (5,1), (5,3), (8,2); diamond;

8. (3,5), (5,0), (5,10), (7,5); diamond.

9. (4,3), (6,1), (6,5), (8,3); both.

10. (7,2), (8,0), (8,4), (9,2); diamond.

3.5.1   QUADRATICS THROUGH THREE POINTS

Points: (0,3), (1,2), (2,3)
Solution: $f(x) = x^2 - 2x + 3$

Points: (0,1), (1,1), (2,3)
Solution: $f(x) = x^2 - x + 1$

Points: (0,1), (1,3), (2,4)
Solution: $f(x) = -x^2 + 3x + 1$

Points: (0,-2), (1,1), (2,8)
Solution: $f(x) = x^2 - 2x + 3$

Points: (0,3), (1,1), (2,1)
Solution: $f(x) = x^2 - 3x + 3$

Points: (1,0), (2,0), (3,2)
Solution: $f(x) = x^2 - 3x + 2$

Points: (1,3), (2,7), (3,13)
Solution: $f(x) = x^2 + x + 1$

Points: (1,3), (2,11), (3,21)
Solution: $f(x) = x^2 + 5x - 3$

Points: (1,5), (2,6), (3,5)
Solution: $f(x) = -x^2 + 4x + 2$

Points: (1,6), (2,12), (3,22)
Solution: $f(x) = 2x^2 + 4$

Points: (-1,3), (2,7), (5,13)
Solution: $f(x) = x^2 + 1$

Points: (-1,1), (2,7), (3,13)
Solution: $f(x) = x^2 + x + 1$

Points: (-1,5), (2,8), (3,17)
Solution: $f(x) = 2x^2 - x + 2$

Points: (-1,6), (2,3), (3,-2)
Solution: $f(x) = -x^2 + 7$

Points: (-1,-10), (2,11), (3,22)
Solution: $f(x) = x^2 + 6x - 5$

Points: (-2,5), (0,-7), (2,5)
Solution: $f(x) = 3x^2 - 7$

Points: (-2,21), (0,7), (2,9)
Solution: $f(x) = 2x^2 - 3x + 7$

Points: (-2,7), (0,1), (2,3)
Solution: $f(x) = x^2 - x + 1$

Points: (-2,-1), (0,-1), (2,7)
Solution: $f(x) = x^2 + 2x - 1$

Points: (-2,5), (0,5), (2,13)
Solution: $f(x) = x^2 + 2x + 5$

Points: (-3,4), (-2,-1), (-1,-1)
Solution: $f(x) = x^2 + 3x + 1$

3.5.2   STRANGE SQUARES

Shown in Figure 3.7 is a collection of strange squares defined by vectors $\vec{v} = (a, b)$ for $a = b = 2$, $a = 4$, $b = 1$, $a = b = 3$, $a = 3$, $b = 2$, $a = 5$, $b = 2$, and $a = b = 4$.

3.5.3   AREAS ON A GRID

For each of the shaded areas below, find the shaded area and give a reason your answer is correct in terms of simple area formulas.

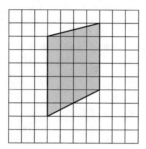

Solution: This shape can be decomposed into an upper triangle, base 4, height 1, 4×4 square, lower triangle width 4 height 2. $A = \frac{1}{2}4 \cdot 1 + 4 \cdot 4 + \frac{1}{2}4 \cdot 2 = 2 + 16 + 4 = 22$.

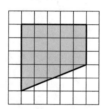

Solution: This shape can be decomposed into a 5×3 rectangle and a triangle with base 5 and height 2 yielding an area of $A = 5 \cdot 3 + \frac{1}{2}5 \cdot 2 = 15 + 5 = 20$.

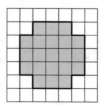

Solution: This shape is a 5×5 square with four 1×1 squares removed for an area of $A = 5^2 - 4 \cdot 1^1 = 21$.

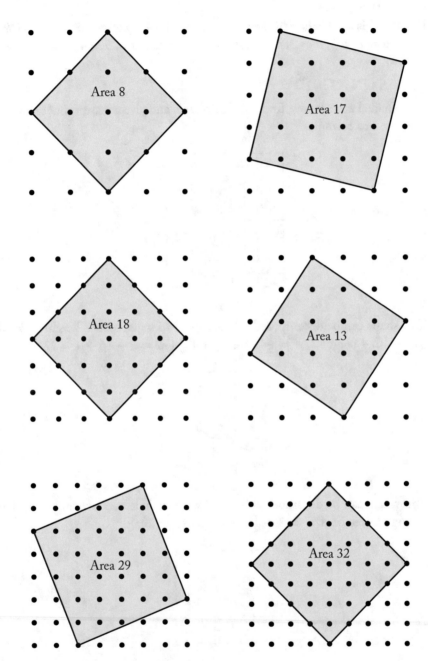

Figure 3.7: A collection of squares defined by vectors.

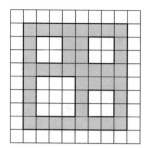

Solution: The large square is 8×8. We need to reduce the area by one 3×3 square and three 2×2 squares. This totals $A = 64 - 9 - 3 \cdot 4 = 41$.

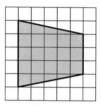

Solution: This area decomposes into two triangles with base 5 and height 1 and a 3×5 rectangle. This gives us $A = 2 \times \frac{1}{2} \cdot 5 \cdot 1 + 3 \cdot 5 = 5 + 15 = 20$.

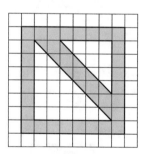

Solution: The large square is 8×8. We need to reduce the area by half a 6×6 and half a 4×4 square. This gives us $A = 64 - (\frac{1}{2}36) - (\frac{1}{2}16) = 64 - 18 - 8 = 38$.

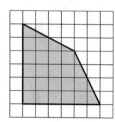

Solution: This shape can be viewed as a 6×6 square with a 2×2 square and two triangles with base 4 and height 2 removed. This gives us $A = 36 - 4 - 2 \cdot \frac{1}{2} \cdot 4 \cdot 2 = 24$.

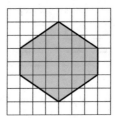

Solution: This shape breaks up as a 6×6 square with four triangles removed that have base 3 and height 2. This gives us $A = 36 - 4 \cdot \frac{1}{2} \cdot 3 \cdot 2 = 36 - 12 = 24$.

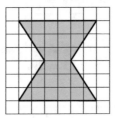

Solution: This shape breaks into a 2×6 rectangle and four triangles with a base of 2 and a height of 3. This gives us area $A = 2 \times 6 + 4 \cdot \frac{1}{2} \cdot 2 \cdot 3 = 24$.

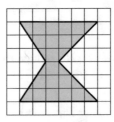

Solution: This shape breaks into a 1×6 rectangle, two triangles with a base of 2 and a height of 3, and two triangles with a height and base of 3. This gives us area $A = 1 \times 6 + 2 \cdot \frac{1}{2} \cdot 2 \cdot 3 + 2 \cdot \frac{1}{2} \cdot 3 \cdot 3 = 21$.

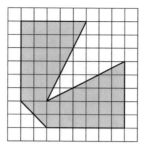

Solution: This shape breaks into half of a 2×2 square, two 2×6 rectangles, and two triangles with base and height 3,6 and 6,3. Treating the two complementary triangles as a single 6×3 rectangle we get $A = 4 - (\frac{1}{2}2) \cdot 2 + 2 \cdot 2 \cdot 6 + 3 \cdot 6 = 44$.

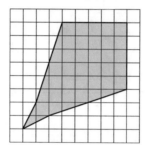

Solution: This shape can be viewed as an 8×8 square with two triangles with base and height 1,2 and 2,1 removed, two 1×6 rectangles removed, and two triangles with base and height 2,6 and 6,2 removed. Combining the complementary triangles into a 2×1 rectangle and a 2×6 rectangle, this gives us $A = 64 - 2 \cdot 1 - 2 \cdot 6 \cdot 1 - 2 \cdot 6 = 64 - 2 - 12 - 12 = 38$.

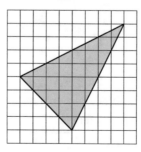

Solution: This shape can be viewed as an 8×8 square with three triangles removed. They have base and height 4,4; 4,8; 8,4. This gives us an area of $A = 64 - \frac{1}{2}(16 + 32 + 32) = 64 - 40 = 24$.

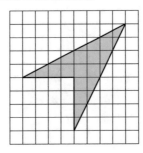

Solution: This shape is that in the previous problem with a triangle with base and height 4 removed. This gives us an area of $A = 24 - (\frac{1}{2}4) \cdot 4 = 24 - 8 = 16$.

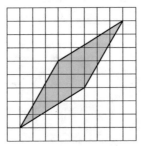

Solution: This shape can be viewed as an 8×8 rectangle with four different triangles with base and height 3,5 and 5,3 removed. This gives us an area of $A = 64 - 4 \cdot \frac{1}{2} \cdot 3 \cdot 5 = 64 - 30 = 34$.

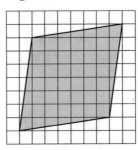

Solution: This shape can be viewed as an 8×8 rectangle with four different triangles with base and height 1,7 and 7,1 removed. This gives us an area of $A = 64 - 4 \cdot \frac{1}{2} \cdot 1 \cdot 7 = 64 - 14 = 50$.

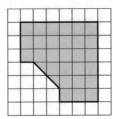

Solution: This is a shape that is a 6×6 square, less a 3×3 square with half a 2×2 square put back. This gives us an area of $A = 36 - 9 + \frac{1}{2} \cdot 4 = 29$.

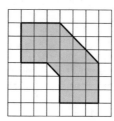

Solution: This is a shape that is a 6×6 square, less a 3×3 square with half a 1×1 square put back and half of a 3 square removed. This gives us an area of $A = 36 - 9 + \frac{1}{2} \cdot 1 - \frac{1}{2} \cdot 9 = 23$.

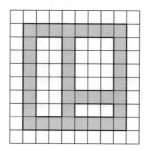

Solution: This shape is an 8×8 square less rectangles with dimensions 2×6, 3×4, and 3×1 giving us an area of $A = 64 - 12 - 12 - 3 = 37$.

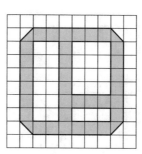

Solution: This shape is the shape from the previous problem with four halves of a 1×1 square removed. This yields an area of $A = 37 - 4 \times 12 = 35$.

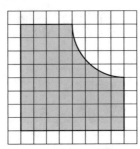

Solution: This shape is an 8×8 square with one-fourth of a circle of radius four removed from it. This gives us an area of $A = 8 \cdot 8 - \frac{1}{4}\pi 4^2 = 64 - 4\pi \cong 51.43$.

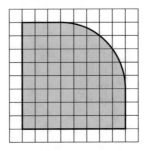

Solution: This shape starts as an 8×8 square, removes the upper-rightmost 5×5 square, and then adds back a radius 5 quarter circle. This gives us an area of $A = 8 \cdot 8 - 5 \cdot 5 + \pi \cdot 5^2 \cong 58.63$.

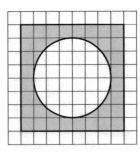

Solution: This shape starts as an 8×8 square and has a circle of radius three removed from it. This gives us an area of $A = 64 - \pi \cdot 3^2 = 64 - 9\pi \cong 35.73$.

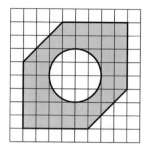

Solution: This shape starts as an 8×8 square and has two halves of a 3×3 square a circle of radius two removed from it, yielding an area of $A = 64 - 9 - \pi \cdot 2^2 = 55 - 4\pi \cong 42.43$.

CHAPTER 4

Problems Using Whole Numbers

4.1   SUMS OF CONSECUTIVE INTEGERS

Here is a question that seems very simple to start, but it turns out to have a rich depth for a student to explore: if I give you a positive numbers can you write it as a sum of consecutive integers?

The answer is a qualified yes for every positive whole number, and there is some elementary mathematics involved in showing why. The qualification comes from the fact that "a sum of consecutive integers" does not specify *how many* integers should be involved in that sum. This is partially why this is a rich problem factory. Some numbers have more than one solution, and if you really want to challenge your students, you could phrase the question as "for a given positive whole number, find *every* sum of consecutive integers that adds up to that whole number."

There is an obvious answer to each problem factory, and that is if the whole number is n, then $n = n$ technically does work, although we call that the *trivial* case in mathematics for good reason. You may want to clarify for your students that you want to see a sum of more than one integer for a given whole number.

It is worth noting that some of the sums have negative terms, as we see in the following example.

Example 4.1   Let $n = 9$. Then we can see that the sequences adding up to 9 are:

$9 = 9$
$9 = 2 + 3 + 4$
$9 = -3 - 2 - 1 + 0 + 1 + 2 + 3 + 4 + 5$
$9 = 4 + 5$
$9 = -1 + 0 + 1 + 2 + 3 + 4$
$9 = -8 - 7 - 6 - 5 - 4 - 3 - 2 - 1 + 0 + 1 + 2 + 3 + 4 + 5 + 6 + 7 + 8 + 9.$

You could also restrict the problem to sums of positive whole numbers, in which case your students could discover that there are some positive whole numbers that cannot be the sum of more than 1 consecutive positive whole integers. This problem factory is easy to understand, and the implementation is immediate, as picking a number takes nearly no time at all and having students try to find solutions for a series of numbers should give them insight into the general approach and the formulas given below.

As part of our service for problem factories, we worked out some facts that govern what non-trivial solutions to this sort of problem look like. Here are two useful facts about consecutive sum problems.

- If n can be written as the sum of an odd number of consecutive terms, then the number of terms in the sum is a divisor d of n and the center term of that sum n/d.

- If n can be written as the sum of an even number of consecutive terms, then half the number of terms in the sum is a divisor d of n and the middle two terms of the sum add to n/d, which must be odd.

If we wanted numbers that add to $n = 25$, for example, then there is a sum of 5 terms with the middle term equal to $25/5 = 5$. This is $25 = 3 + 4 + 5 + 6 + 7$. To get a sum of an even number of terms that add to $n = 25$ we choose the divisor $d = 5$ again. The second fact tells us that ten terms are required and that the middle two terms add to $25/5 = 5$, making those terms 2 and 3. This gives us the sum $25 = -2 - 1 + 0 + 1 + 2 + 3 + 4 + 5 + 6 + 7$. These two facts can be used to generate many consecutive sum problems.

4.2   PIZZA COUNTING

This problem factory is intended to introduce simple fundamental principles of counting up to and including binomial coefficients. There are two principals that underlie this sort of counting. The task is to decide, given a number of choices that can be made when ordering a pizza, how many different pizzas one could order. There are two fundamental counting results we need to start with.

If two choices have the property that the outcome of each places no constraint on the other then we say the choices are **independent**. In terms of pizzas, ordering pepperoni doesn't affect your choice of kind of cheese, for example.

The number of pizzas you could order is based on how many choices you have, and how many options per choice. The total number of pizzas is the product of the number of options for choice 1 multiplied by the number of options for choice 2 multiplied by the number of options for choice 3 and so on.

This result is often summarized with the phrase *independent choices multiply*.

A simple example. Suppose there are nine available veggie toppings and six available meat toppings available for a pizza. How many pizzas are there with one veggie and one meat topping? Since which veggie we chose does not affect which meat be may choose, the answer is a very simple $9 \times 6 = 54$.

If option 1 for choice 1 has the property that it prevents us from choosing option 1 for choice 2, then we say those options are **mutually exclusive**. If we are ordering either a one-topping pizza or a two-topping pizza then those two choices are mutually exclusive; the number of toppings is one or two, not both.

If a set of options falls into mutually exclusive groups then the total number of pizzas is the sum of the number of options in each of those groups.

This result is often summarized with the phrase *mutually exclusive choices add*.

Let's work an example. Suppose we have 12 available toppings. How many ways are there to order a one-topping or a two-topping pizza if the toppings on the two topping pizza must be different?

Answer: First of all, there are 12 one-topping pizzas, so the answer will be 12 plus the number of two-topping pizzas. The wrinkle that the toppings must be different means that there are 12 choices for the first topping and 11 left for the second, which looks like $12 \times 11 = 132$ two-topping pizzas. There is a problem there: sausage and mushroom is the same pizza as mushroom and sausage. Since we do not care which topping is first or second, the count of 132 includes each two-topping pizza *twice*. This means that the number of two-topping pizzas, for 12 possible toppings, is $132/2 = 66$. We can now get the answer: $12 + 66 = 78$ possible one- or two-topping pizzas.

Suppose we went on to three-topping pizzas. We would over-count the three-topping pizzas the way we over-counted the two-topping pizzas, but it would be worse, because there are six possible ways to order the same pizza with three toppings. In fact, there is a general way to work with this over-counting problem, which we call *binomial pizza*.

4.2.1   BINOMIAL PIZZA

Suppose that we have n meat toppings and m veggie toppings. How many one-meat, one-veggie pizzas are there? Using the idea: *independent choices multiply*, the answer is $n \times m$. This situation is made simple by the independence of the choice of the meat and the veggie topping.

What if we have two meat toppings? Here we hit a problem with the definition of "two toppings": is it a double application of one topping a double topping? You can attack this this way. A pizza with two toppings has either a double topping of one type or two different toppings. The doubled topping pizzas have only one type of topping so, if there are n meat toppings there are n of these double topping pizzas.

What about the pizzas with two different toppings? There is a problem here: is pepperoni and sausage different from sausage and pepperoni? Short of forensic examination, it's hard to tell which topping was applied first, so we will go with the answer "no." That means if we take n meat toppings as independent choices we will get each pizza with two different toppings twice. Once we know this it is not too difficult to count the number of two meat topping pizzas with distinct toppings. There are n^2 two meat choices of which n are doubled single toppings. That means the pizzas with two different toppings can be counted as $n^2 - n$ by noting that "two different toppings" and "the same topping twice" are mutually exclusive possibilities. Now, since we know these two topping pizzas each appear twice, we divide by two and get that there are

$$\frac{1}{2}(n^2 - n)$$

pizzas with two different meat toppings. This means the total number of two topping meat pizzas are $\frac{1}{2}(n^2 - n) + n$ or $\frac{1}{2}(n^2 + n)$.

Readers already familiar with **binomial coefficients** will have noticed that the number of pizzas with two different meat toppings can already be counted as the number of ways to choose two of n objects, the binomial coefficient $\binom{n}{2} = \frac{1}{2}n(n-1)$. Using binomial coefficients makes choosing pizza toppings easier—as least as long as they are different. If you are not familiar with binomial coefficients, take a look at the **Delving Deeper** box.

Here is another example. Suppose we want to order two pizzas, each with two toppings, and at least one of the pizzas is vegetarian. This is a little tricky. First, we look at the mutually exclusive cases: one veggie pizza or two veggie pizzas. If there is one veggie pizza that case splits in two as well into one or two meat toppings on the non-veggie pizza. This means we can do three cases and then add them up. Let's start with the one two-veggie pizza and one two-meat pizza. This is

$$\frac{1}{2}n(n+1)\frac{1}{2}m(m+1) = \frac{n(n+1)m(m+1)}{4},$$

where n is the number of veggie toppings and m is the number of meat toppings, and the choices of veggie and meat ingredients are independent. Notice we are re-using our result from the last example that a two-topping pizza has $\frac{1}{2}n(n+1)$ outcomes, counting a pizza with a double topping.

The next case is one two-veggie and one veggie and meat pizza. We chose the veggie pizza $\frac{1}{2}n(n+1)$ ways, again re-using our result for a pizza with two ingredients of the same type. We then chose a meat and a veggie ingredient $n \times m$ ways and multiply, since the choices are independent, getting a $\frac{1}{2}n^2(n+1)m$ possibilities.

If both the pizzas are veggie then each pizza can be chosen independently of the other giving us $\left(\frac{1}{2}n(n+1)\right)^2 = \frac{1}{4}n^2(n+1)^2$ possibilities. If we add up all three cases then the number of possible pizza orders is:

$$\frac{1}{4}nm(n+1)(m+1) + \frac{1}{2}n^2(n+1)m + \frac{1}{4}n^2(n+1)^2$$

$$= \frac{(n+1)(m+1)}{4}\left[1+n^2\right],$$

which is a lot of pizza orders. Suppose we go back to an earlier example with nine veggie and six meat toppings. This would come to 5,400 possible ways to order the two pizzas.

$$
\begin{array}{ccccccccccccccccccc}
&&&&&&&&& 1 \\
&&&&&&&& 1 && 1 \\
&&&&&&& 1 && 2 && 1 \\
&&&&&& 1 && 3 && 3 && 1 \\
&&&&& 1 && 4 && 6 && 4 && 1 \\
&&&& 1 && 5 && 10 && 10 && 5 && 1 \\
&&& 1 && 6 && 15 && 20 && 15 && 6 && 1 \\
&& 1 && 7 && 21 && 35 && 35 && 21 && 7 && 1 \\
& 1 && 8 && 28 && 56 && 70 && 56 && 28 && 8 && 1 \\
1 && 9 && 36 && 84 && 126 && 126 && 84 && 36 && 9 && 1 \\
\end{array}
$$

$$1 \quad 10 \quad 45 \quad 120 \quad 210 \quad 252 \quad 210 \quad 120 \quad 45 \quad 10 \quad 1$$

Figure 4.1: The first 10 rows of Pascal's Triangle.

Delving Deeper: Binomial Coefficients and Pascal's Triangle

Recall that a *binomial* is a polynomial of the form $(x + y)$, where x and y are variables. If we raise a binomial to some positive integer power, n, so that it takes the form $(x + y)^n$, we can expand that expression into $n + 1$ terms. For example, with $(x + y)^2$, we get $x^2 + 2xy + y^2$. A really interesting phenomenon accompanies this process: for any value of n, the coefficients of the variable products are the values in the nth row of Pascal's Triangle! See Figure 4.1.

When expanding any binomial of the form $(x + y)^n$, it always takes the form

$$
(x + y)^n = \binom{n}{0}x^n y^0 + \binom{n}{1}x^{n-1}y^1 + \cdots + \binom{n}{n-1}x^1 y^{n-1} + \binom{n}{n}x^0 y^n,
$$

where we read $\binom{n}{k}$ as the nth row of Pascal's Triangle, and k as the kth number in the row reading from left to right, and we start counting at $k = 0$. For example,

$$
(x + y)^4 = \binom{4}{0}x^4 y^0 + \binom{4}{1}x^3 y^1 + \binom{4}{2}x^2 y^2 + \binom{4}{3}x^1 y^3 + \binom{4}{4}x^0 y^4
$$

$$
(x + 4)^4 = 1 \cdot x^4 + 4 \cdot x^3 y + 6 \cdot x^2 y^2 + 4 \cdot xy^3 + 1 \cdot y^4.
$$

Here is the shortcut for choosing distinct pizza ingredients. If $0 \leq m \leq n$ then the number of ways to choose m different objects from a set of n objects is

$$\binom{n}{m} = \frac{n!}{m!(n=m)!},$$

where $n!$ is called a **factorial**, and $n! = n \times (n-1) \times (n-2) \times \ldots \times 3 \times 2 \times 1$. These numbers appear in *Pascal's triangle* and count choices.

So, for example, if we have seven meat toppings available, then the number of pizzas with three different meat toppings we can order is

$$\binom{7}{3} = \frac{7!}{3! \cdot 4!} = \frac{5040}{6 \cdot 24} = 35.$$

What if we have any three meat toppings, including repeats? Assuming we have n meat toppings then there are n triple meat pizzas with only one type of topping. We also know there are

$$\binom{n}{3} = \frac{1}{6}(n)(n-1)(n-2)$$

pizzas with three different toppings. What about the ones with one topping doubled and a single dose of another? For those we choose the double topping, it is one of n possibilities, and the second, single topping from the $n-1$ remaining choices. This is a lot like the argument for choosing two toppings *except* that the fact one of the toppings is doubled means we are not double counting the possibilities. So we get $n(n-1) = n^2 - n$ three-topping pizzas with one of the toppings doubled up. Totaling this we see there are

$$n + n(n-1) + \frac{1}{6}n(n-1)(n-2) = \frac{1}{6}n(n+1)(n+2)$$

ways to put three of n meat toppings on a pizza with any pattern of repeats.

Since the meat and veggie toppings are chosen independently, it is possible to pose problems with and without doubling up meat and veggie choices to generate a huge number of problems. You can also pose questions like "Mike's Pizzeria says they have over one million pizzas you can order. Is this true?" Finally, if there are not yet enough problems, you have have double cheese, different types of cheese, crust, and sauce. The individual categories use the same mechanics as the meat toppings, though two types of crust may be a bit silly.

Here is another example of a pizza question. Imagine that you and your friends want to order a pizza. The pizza place has a deal where you can get 3 of 7 meat toppings and 3 of 7 veggie toppings. So how many different pizzas could you order?

The number of meat choices is $\binom{7}{3} = 35$ and the number of veggie choices is $\binom{7}{3} = 35$, and since those choices are independent, we can multiply them. So the total number of pizzas you could order is $35^2 = 1225$.

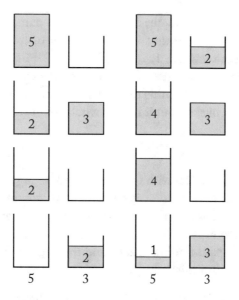

Figure 4.2: This is a solution, showing eight steps, to the problem of measuring a single liter of water using only three- and five-liter containers. The solutions are read from top to bottom, left column first. So 5 and 0 is the first step, and 5 and 2 is the fifth step.

4.3   FIBONACCI JARS

Suppose that we have two jars and a large pool of water. The first holds 5 liters of water, the second holds three. Can we measure out 1 liter of water? You can fill a jar, empty it out, or pour the contents of one jar into another. You cannot partially fill a jar to get a measured amount of water. Figure 4.2 shows a solution to the problem of measuring 1 liter of water with a 3- and a 5-liter container.

This problem could have been posed for any two bucket sizes and target amounts of water. To make this a problem factory, we need to know the pattern of buckets and targets that will work. The amounts of water it is possible to measure out are those that are multiples of the greatest common divisor of the bucket sizes. If the amount you want to measure is larger than either bucket, then you need a third container into which you can pour the water being measured.

It is also easily possible to construct *impossible* problems. If the target amount of water is *not* an even multiple of the greatest common divisor of the bucket sizes, the problem cannot be done. The most obvious version of this is when the two bucket sizes are both even and the target amount of water is odd—students just learning their number sense could benefit from this sort of problem. If you have a 12-ounce cup and a 15-ounce cup and have a target of 7 ounces, then it is less obvious that the problem is impossible. The number 7 is not a multiple of 3, which is the greatest common divisor of 12 and 15; this will take students a bit longer to notice.

Let's wind up by answering the question, why is the name of this problem "Fibonacci Jars." When you are computing a greatest common divisor of two numbers that are adjacent Fibonacci numbers, from the sequence 1, 1, 2, 3, 5, 8, 13, 21, 34,... these pairs take the most steps to measure 1 liter of water, because of the number of steps involved in computing their greatest common divisor. For example, this means that you *can* measure 1 liter of water using a 21-liter bucket and a 13-liter bucket, but it takes a whole lot of steps.

4.4 ADVANCE BY FACTORS

An important part of learning to deal with fractions is generating a sense of when a fraction is in simplest form. This means that, as soon as students start working with fractions, a sense of how whole numbers factor is critical. This, in turn, makes games and puzzles that have to do with factoring important resources.

Advance by Factors is a path finding puzzle, and the path from one number to another is created by knowing the factors of numbers. The puzzle works as follows.

1. The instructor or the student picks a goal number that the student is going to try to reach in the shortest number of moves, say 48, for example.

2. The student begins the puzzle at the number 2.

3. A student can only move from their current number using one of the factors of that number. The student then adds one of the factors to move to another number. For example, if the student starts at 2, then the factors are 1 and 2. If the student adds 1 to 2, now the student is at 3. If the student adds 2 to 2, then the student is now at 4.

4. If the current number is prime, the student may use any factor of the current number to advance to another number.

5. If the current number is not prime, the player must use a factor strictly between 1 and the current number. If the current number was 9, for example, the factors of 9 are 1, 3, and 9, the only move available is to add 3 to 9 and obtain 12.

6. The goal is to reach the target number, 48 in our current example, in the shortest number of moves.

A map of moves to 48, is shown in Figure 4.3. One of the shortest paths is 2, 4, 6, 9, 12, 18, 24, 32, 48.

4.4.1 VARIATIONS

Advance by Factors is a wonderfully simple problem factory to construct very quickly. The goal number and the starting number may both be changed.

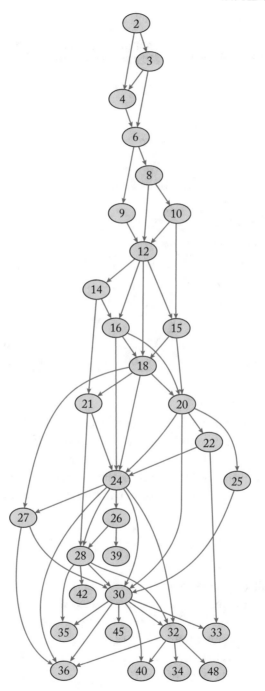

Figure 4.3: A map of moves in *advance by factors* to reach 48.

Table 4.1: Minimum number of moves to reach 243

Moves	Target
0	2
1	3 4
2	6
3	8 9
4	10 12
5	14 15 16 18
6	20 21 24 27
7	22 25 26 28 30 32 36
8	33 34 35 38 39 40 42 45 **48** 54
9	44 49 50 51 52 56 57 60 63 64 72 81
10	46 55 58 62 65 66 68 70 74 75 76 78 80 84 90 96 108
11	69 77 82 85 86 87 88 91 92 93 95 98 99 100 102 104 105 110 111 112 114 117 120 126 128 135 144 162
12	94 106 115 116 119 121 122 123 124 125 129 130 132 133 136 138 140 146 147 148 150 152 153 156 160 164 165 168 171 180 189 192 216 243

One thing to note is that it is possible to obtain a value for the current number that makes it impossible to reach the target number. This is a natural part of the investigation process, just have the student return to the start and try another path. For your students who are a little advanced, give them a few prime number targets. They should quickly figure out that any prime number other than 2 and 3 cannot be reached playing this game, assuming you start at 2.

4.4.2 LENGTHS OF SHORTEST PATHS FOR NUMBERS OF MOVES TO A TARGET

We used a technique called *dynamic programming* to find the minimum number of moves required to reach a given target number in advance by factors, up to 243. Table 4.1 contains all the numbers that can be reached in 12 or fewer moves starting with 2. The only numbers that cannot be reached *at all* are prime numbers bigger than 3.

We hope that Table 4.1 is helpful for an instructor giving this problem factory to a student; it is the key for a correct number of steps in an answer.

4.5   NUMBER SENTENCE PUZZLES

A *number sentence* is a way of sneaking up on the idea of an equation. So

$$5 * (2 + 3) = 25$$

is a number sentence—it is just not a very interesting one. A number sentence is interesting if the same digits or numbers occur in it fairly often. If we remember that $3 \times 37 = 111$ then it is a short jump to the really interesting number sentence

$$3 \times 7 \times 37 = 777$$

which only uses two digits. There aren't a lot of number sentences like that around.

We offer three different number sentence puzzles.

The first asks a student to fill in the digits, so the example with all the threes and sevens would work here:

$$\square \times \diamond \times \square \diamond = \diamond \diamond \diamond$$

All the spaces represented by the same symbol must use the same digit.

The second sort of number sentence puzzle gives the student a sentence with blanks and a list of digits. The digits are used (once each) to make the sentence true. If the number sentence is

$$(\square + \square) \times (\square\square - \square) = \square\square$$

and the list of digits was 1, 1, 2, 3, 4, 7, 8 then a solution would be:

$$(1 + 2) \times (34 - 7) = 81.$$

The third sort of number sentence puzzle is similar to the first, but one side of the equation is a **?** representing a goal. The students are supposed to fill in the digits to make the value represented by the **?** as large as possible. For example:

$$(\square + \square) \times (\square + \square) \times (\square + \square) \times = ?$$

with the digits being 1, 2, 3, 4, 5, 6 has a maximum value of 343 for **?**.

It is possible to create problems of this sort that are easy to solve because there are few options, e.g.,

$$\square \times (\square\square + \square) = ?$$

with digits 5, 5, 5, 1. This one is pretty obvious—put the one in front and you waste a lot of potential and only get 60. There are two solutions:

$$5(51 + 5) = 5(55 + 1) = 280.$$

It does not matter which of the one's digits is filled with the one and you also clearly do not want to use the one for the tens digit inside the parenthesis. This sort of simple problem can be good for introducing these problems and building confidence.

4.6   EXAMPLE PROBLEMS AND PUZZLES

4.6.1   SUM OF CONSECUTIVE NUMBER PROBLEMS

If we leave out the one-term solution "9", how many ways can we write 9 as the sum of positive, consecutive whole numbers?
Answer: two; $2 + 3 + 4$, $4 + 5$.

Can you write 11 as a sum of 22 consecutive whole numbers?
Answer: yes; $-10 - 9 - 8 - \cdots + 0 + 1 + 2 + 3 + \cdots + 10 + 11$. All the numbers *except 11* are either 0 or can be paired with their negative.

Can you write 15 as a sum of 3 consecutive whole numbers?
Answer: yes; $4 + 5 + 6 = 15$.

Can you write 15 as a sum of 4 consecutive whole numbers?
Answer: no; it is not possible, the sum of four consecutive whole numbers must be even.

Can you write 15 as a sum of 5 consecutive whole numbers?
Answer: yes; $1 + 2 + 3 + 4 + 5 = 15$.

Can you write 25 as the sum of 5 consecutive terms?
Answer: $3 + 4 + 5 + 6 + 7$.

Can you write 27 as the sum of 3 consecutive whole numbers?
Answer: yes; $8 + 9 + 10 = 27$.

Can you write 27 as the sum of 6 consecutive whole numbers?
Answer: yes; $2 + 3 + 4 + 5 + 6 + 7 = 27$.

Find all the ways to write 30 as the sum of positive, consecutive terms.
Answer: 30 or $9 + 10 + 11$ or $6 + 7 + 8 + 9$ or $4 + 5 + 6 + 7 + 8$. Other sums involve negative terms.

Find all the ways to write 35 as the sum of positive, consecutive terms.
Answer: 35 or $5 + 6 + 7 + 8 + 9$ or $2 + 3 + 4 + 5 + 6 + 7 + 8$.

Find a way to write 105 as the sum of 4 or more positive consecutive whole numbers.
Answer: $19 + 20 + 21 + 22 + 23$ or $12 + 13 + 14 + 15 + 16 + 17 + 18$ or $15 + 16 + 17 + 18 + 19 + 20$ or $6 + 7 + 8 + 9 + 10 + 11 + 12 + 13 + 14 + 15$; there are others but these are the shorter ones.

4.6.2   PIZZA MATH

Suppose we have 12 toppings available. How many two-topping pizzas are possible?
Answer: $\binom{12}{2} = 66$

Suppose we have 12 toppings available. How many pizzas are possible with no more than two toppings?

Answer: There can be zero, one, or two toppings so the answer is $\binom{12}{0} + \binom{12}{1} + \binom{12}{2} = 1 + 12 + 66 = 79$. The zero-topping pizza is a cheese pizza.

How many different kinds of pizzas can we order if we can choose 2 of 5 meat toppings, 2 of 6 veggie toppings, and 2 of 3 kinds of cheeses?

Answer: $\binom{5}{2} \times \binom{6}{2} \times \binom{3}{2} = 450$ different kinds of pizzas.

Suppose the Giancarlo's Pizzaria offers 5 cheese blends, 3 sorts of crust, 4 possible sauces, 22 meat toppings, and 31 veggie toppings. How many one-topping pizzas can you order? How many pizzas with one meat and one veggie topping can you order?

Answer: One topping—$5 \times 3 \times 4 \times (22 + 31) = 60 \times 53 = 3180$.
Two toppings—$5 \times 3 \times 4 \times 22 \times 31) = 40,920$.

How many different kinds of pizzas can we order if: we are ordering two pizzas and on the first pizza we are choosing 1 of 5 meat toppings and 3 of 5 veggie toppings, and on the second pizza we are choosing 3 of 6 meat toppings and 2 of 4 veggie toppings?

Answer: $\binom{5}{1} \times \binom{5}{3} \times \binom{6}{3} \times \binom{4}{2} = 6000$ different combinations of two pizzas.

How many pizzas can we order if we are ordering a pizza with the following rules: we can have either meat toppings or veggie toppings, but not both. We can choose 3 from 6 meat toppings or 4 from 7 veggie toppings.

Answer: $\binom{6}{3} + \binom{7}{4} = 20 + 35 = 55$ different possible pizzas.

Suppose there are ten meat toppings, twelve veggie toppings, and you want a three topping pizza, with at least one meat topping. How many possibilities are there?

Answer: Segregate the answer by the number of meat toppings, one, two, or three. These are separate cases so we can count them individually and add up the results:

$$\binom{10}{1}\binom{12}{2} + \binom{10}{2}\binom{12}{1} + \binom{10}{3} = 10 \times 66 + 45 \times 12 + 120 = 660 + 40 + 120 = 1320.$$

4.6.3 PROBLEMS ON FIBONACCI JARS

Using an 800-ml and a 500-ml beaker, measure out 300 ml of water.

Answer: Fill the 800; fill the 500 from the 800; there are 300 ml left in the 800.

With a 7-liter jar and a 5-liter jar, devise a way to measure out 3 liters of water.

Answer: Fill the 5-liter jar, empty it into the 7-liter jar. Fill the 5-liter jar again and use it to fill the 7-liter jar. This requires 2 liters, leaving 3 in the 5-liter jar.

With a 7-liter jar and a 5-liter jar, devise a way to measure 8 liters of water into an available 20-liter basin.

Answer: Fill and empty the 5-liter jug into the basin three times. It now contains 15 liters of water. Fill the 7-liter jar from the basin, leaving 8 liters of water behind. The fundamental solution is $3 \times 5 - 7 = 8$.

Using a 4-ounce cup and a 3-ounce cup, measure out 2 ounces of water.

Answer: Fill 4; fill 3 from 4 leaving one ounce. Empty the 3 and transfer one ounce to the 3. Fill 4. Fill 3 from 4 leaving two in the 4. Done.

Using a 13-scruple dish and a 5-scruple thimble, measure out 4 scruples of aqua regia. Assume you have a sufficient source of this substance.

Answer: the fundamental solution is $3 \times 13 - 7 \times 5 = 4$ requiring many steps—a hard problem.

If we are measuring sand with a 1.1-liter scoop and a 0.5-liter scoop then, since $1.1 - 2 \times 0.5 = 0.1$ we should be able to measure out and multiple of one-tenth of a liter of sand, using an auxiliary basin. Find the smallest number of scoops you can to measure out 16.7 liters of sand.

Answer: $12 \times 1.1 + 7 \times 0.5 = 16.7$, 19 scoops are required.

With a 4-liter jug and a 12-liter jug, what is the closest to 93 liters that you can measure?

Answer: $12 \times 7 + 2 \times 4 = 92$, within 1.

Find two containers that can measure 3 cups of water, cannot be used to measure 2 cups, and both of which hold an even number of cups larger than three.

Answer: A 9 and a 6-cup measure are one of many possible answers. Fill the 9-cup measure and then use it to fill the 6-cup measure. This leaves 3 cups in the 9-cup measure. Since both containers hold multiples of 3 cups, measuring 2 cups is not possible.

Number Sentence Puzzles

Directions: Fill in digits that make the number sentence true—use the same digit when the symbols are the same.

$(\triangledown + \square\triangledown) \times (\triangledown + \bigcirc\triangledown) = \bigcirc \triangledown \triangledown$
$(0+10)\times(0+80)=800$

$(\triangle + \triangle\triangle) \times (\bigstar + \triangle\triangle) = \square\square\triangle$
$(2+22)\times(1+22)=552$

$(\triangledown + \square\triangledown) \times (\square + \square\square) = \triangledown \bigcirc \square$
$(9+29)\times(2+22)=912$

$(\square + \bigstar\square) \times (\square + \bigcirc\bigcirc) = \bigcirc \bigcirc \square$
$(0+10)\times(0+55)=550$

$(\cap + \cap\spadesuit) \times (\spadesuit + \triangledown\spadesuit) = \cap \cap \spadesuit$
$(8+80)\times(0+10)=880$

$(\clubsuit + \clubsuit\square) \times (\square + \clubsuit\cap) = \square\square\clubsuit$
$(2+29)\times(9+23)=992$

$(\bigstar + \bigstar\bigstar) \times (\bigstar + \bigstar\triangledown) = \square\square\triangledown$
$(1+11)\times(1+18)=228$

$(\diamondsuit + \diamondsuit\heartsuit) \times (\heartsuit + \bigstar\heartsuit) = \heartsuit\bigstar\heartsuit$
$(2+26)\times(6+16)=616$

$(\clubsuit + \heartsuit\clubsuit) \times (\clubsuit + \square\clubsuit) = \triangle\clubsuit\clubsuit$
$(0+40)\times(0+20)=800$

$(\heartsuit + \heartsuit\triangle) \times (\heartsuit + \heartsuit\heartsuit) = \heartsuit\spadesuit\bigstar$
$(1+14)\times(1+11)=180$

$(\spadesuit + \cap\spadesuit) \times (\spadesuit + \spadesuit\spadesuit) = \spadesuit\spadesuit\square\heartsuit$
$(6+86)\times(6+66)=6624$

$(\bigstar + \bigcirc\spadesuit) \times (\bigcirc + \bigcirc\bigstar) = \bigcirc\bigstar\bigcirc\bigstar$

$(2+90)\times(9+92)=9292$

$(\heartsuit + \square\heartsuit) \times (\square + \heartsuit\heartsuit) = \square\clubsuit\heartsuit\square$
$(7+47)\times(4+77)=4374$

$(\diamondsuit + \cap\diamondsuit) \times (\cap + \cap\square) = \cap\square \cap \cap$
$(6+86)\times(8+81)=8188$

$(\bigcirc + \bigcirc\triangledown) \times (\bigcirc + \clubsuit\clubsuit) = \bigcirc \bigcirc \triangledown \triangledown$
$(1+10)\times(1+99)=1100$

$(\clubsuit + \cap\cap) \times (\clubsuit + \clubsuit\diamondsuit) = \clubsuit\diamondsuit\clubsuit\cap$
$(4+88)\times(4+40)=4048$

$(\square + \bigcirc\clubsuit) \times (\square + \square\bigcirc) = \bigcirc\square \bigcirc \clubsuit$
$(8+94)\times(8+89)=9894$

$(\cap + \spadesuit\cap) \times (\spadesuit + \triangle\spadesuit) = \cap \triangle \triangle\cap$
$(2+42)\times(4+54)=2552$

$(\clubsuit + \heartsuit\bigcirc) \times (\bigstar + \heartsuit\bigcirc) = \heartsuit\heartsuit\heartsuit\heartsuit$
$(7+94)\times(5+94)=9999$

$(\spadesuit + \bigcirc\square) \times (\square + \square\square) = \square\square\spadesuit\triangledown$
$(3+89)\times(9+99)=9936$

$\heartsuit \times (\spadesuit\heartsuit + \clubsuit) = \clubsuit\clubsuit\heartsuit$
$4\times(54+2)=224$

$\triangledown \times (\triangledown \triangledown + \triangledown) = \bigstar \triangle \triangledown$
$8\times(88+8)=768$

$\heartsuit \times (\triangledown \triangledown + \triangledown) = \diamondsuit \triangledown \heartsuit$
$4\times(88+8)=384$

$\triangledown \times (\triangle \triangle + \spadesuit) = \spadesuit \triangle \triangledown$
$8\times(66+5)=568$

$\cap \times (\bigstar\clubsuit + \cap) = \bigstar\clubsuit\cap$
$9\times(72+9)=729$

$\triangle \times (\triangle \triangledown +\triangle) = \heartsuit\heartsuit\heartsuit$
$6\times(68+6)=444$

$\cap \times (\bigstar\clubsuit + \cap) = \bigstar\clubsuit\cap$
$9\times(72+9)=729$

$\cap \times (\triangledown \cap +\cap) = \triangledown \triangledown \clubsuit$
$9\times(89+9)=882$

$\triangledown \times (\bigcirc \triangledown +\triangledown) = \clubsuit\square\triangledown$
$8\times(18+8)=208$

$\triangledown \times (\triangle \triangledown +\triangledown) = \triangle\square\triangledown$
$8\times(68+8)=608$

$(\bigstar + \triangle) \times (\triangledown + \bigcirc) = \bigcirc \bigcirc \bigstar$
$(7+6)\times(8+1)=117$

$(\triangledown + \triangledown) \times (\triangledown + \square) = \bigcirc\clubsuit\triangledown$
$(8+8)\times(8+0)=128$

$(\cap + \triangledown) \times (\triangledown + \cap) = \clubsuit \triangledown \cap$
$(9+8)\times(8+9)=289$

$(\triangle + \triangle) \times (\triangle + \triangle) = \bigcirc\heartsuit\heartsuit$
$(6+6)\times(6+6)=144$

$(\cap + \spadesuit) \times (\spadesuit + \cap) = \bigcirc \cap \triangle$
$(9+5)\times(5+9)=196$

$(\cap + \cap) \times (\heartsuit + \heartsuit) = \bigcirc\heartsuit\heartsuit$
$(9+9)\times(4+4)=144$

$(\clubsuit + \cap) \times (\cap + \clubsuit) = \bigcirc\clubsuit\bigcirc$
$(2+9)\times(9+2)=121$

$(\bigcirc + \triangledown) \times (\triangle + \triangledown) = \bigcirc\clubsuit\triangle$
$(1+8)\times(6+8)=126$

$(\bigcirc + \triangledown) \times (\spadesuit + \triangledown) = \bigcirc \bigcirc \bigstar$
$(1+8)\times(5+8)=117$

$(\bigcirc + \triangle) \times (\cap + \triangledown) = \bigcirc \bigcirc \cap$
$(1+6)\times(9+8)=119$

Directions: Use the given digits to fill in the boxes to make the number sentence true.

Digits: 2, 1, 1, 1, 1, 1, 1, 0, 0
$(\square + \square\square)(\square + \square\square) = \square\square\square$
$(1+10)\times(1+10)=121$

Digits: 3, 2, 2, 2, 2, 2, 2, 1, 1
$(\square + \square\square)(\square + \square\square) = \square\square\square$
$(2+21)\times(2+12)=322$

Digits: 5, 5, 2, 2, 2, 2, 2, 2, 1
$(\square + \square\square)(\square + \square\square) = \square\square\square$
$(2+22)\times(1+22)=552$

Digits: 6, 2, 2, 2, 1, 1, 1, 1, 1
$(\square + \square\square)(\square + \square\square) = \square\square\square$
$(1+12)\times(6+11)=221$

Digits: 9, 4, 4, 4, 4, 4, 1, 1, 1
$(\square + \square\square)(\square + \square\square) = \square\square\square$
$(4+19)\times(4+14)=414$

Digits: 4, 4, 4, 4, 1, 0, 0, 0, 0
$(\square + \square\square)(\square + \square\square) = \square\square\square$
$(0+44)\times(0+10)=440$

Digits: 9, 8, 8, 8, 8, 2, 2, 2, 2
$(\square + \square\square)(\square + \square\square) = \square\square\square$

$(2+22)\times(8+29)=888$

Digits: 7, 6, 2, 1, 1, 1, 1, 1, 1
$(\Box + \Box\Box)(\Box + \Box\Box) = \Box\Box\Box$
$(1+11)\times(7+11)=216$

Digits: 8, 4, 1, 1, 1, 1, 1, 1, 1
$(\Box + \Box\Box)(\Box + \Box\Box) = \Box\Box\Box$
$(1+14)\times(1+11)=180$

Digits: 6, 6, 6, 6, 6, 6, 5, 4, 4, 4
$(\Box + \Box\Box)(\Box + \Box\Box) = \Box\Box\Box\Box$
$(6+66)\times(6+56)=4464$

Digits: 9, 5, 4, 4, 4, 4, 4, 4, 4, 2
$(\Box + \Box\Box)(\Box + \Box\Box) = \Box\Box\Box\Box$
$(4+44)\times(4+49)=2544$

Digits: 8, 8, 8, 8, 7, 4, 4, 4, 4, 4
$(\Box + \Box\Box)(\Box + \Box\Box) = \Box\Box\Box\Box$
$(8+84)\times(8+44)=4784$

Digits: 9, 9, 9, 9, 9, 2, 2, 2, 2, 0
$(\Box + \Box\Box)(\Box + \Box\Box) = \Box\Box\Box\Box$
$(2+90)\times(9+92)=9292$

Digits: 9, 9, 9, 9, 9, 8, 8, 8, 8, 3
$(\Box + \Box\Box)(\Box + \Box\Box) = \Box\Box\Box\Box$
$(9+89)\times(3+98)=9898$

Digits: 5, 4, 4, 2, 2, 2, 2, 2, 2, 1
$(\Box + \Box\Box)(\Box + \Box\Box) = \Box\Box\Box\Box$
$(4+22)\times(2+45)=1222$

Digits: 8, 8, 7, 6, 1, 1, 1, 1, 1, 1
$(\Box + \Box\Box)(\Box + \Box\Box) = \Box\Box\Box\Box$
$(8+11)\times(1+61)=1178$

Digits: 8, 8, 8, 8, 8, 8, 2, 2, 1, 0
$(\Box + \Box\Box)(\Box + \Box\Box) = \Box\Box\Box\Box$
$(0+88)\times(8+18)=2288$

Digits: 8, 8, 8, 8, 8, 8, 5, 4, 4, 1
$(\Box + \Box\Box)(\Box + \Box\Box) = \Box\Box\Box\Box$
$(4+85)\times(4+88)=8188$

Digits: 5, 4, 4, 4, 2, 2, 2
$\Box(\Box\Box + \Box) = \Box\Box\Box$
$4\times(54+2)=224$

Digits: 8, 8, 8, 8, 8, 7, 6
$\Box(\Box\Box + \Box) = \Box\Box\Box$
$8\times(88+8)=768$

Digits: 8, 8, 8, 8, 4, 4, 3
$\Box(\Box\Box + \Box) = \Box\Box\Box$
$4\times(88+8)=384$

Digits: 8, 8, 6, 6, 6, 5, 5
$\Box(\Box\Box + \Box) = \Box\Box\Box$
$8\times(66+5)=568$

Digits: 9, 9, 9, 7, 7, 2, 2
$\Box(\Box\Box + \Box) = \Box\Box\Box$
$9\times(72+9)=729$

Digits: 8, 6, 6, 6, 4, 4, 4
$\Box(\Box\Box + \Box) = \Box\Box\Box$
$6\times(68+6)=444$

Digits: 9, 9, 9, 7, 7, 2, 2
$\Box(\Box\Box + \Box) = \Box\Box\Box$
$9\times(72+9)=729$

Digits: 9, 9, 9, 8, 8, 8, 2
$\Box(\Box\Box + \Box) = \Box\Box\Box$

9×(89+9)=882

Digits: 4, 3, 3, 2, 2, 2, 1
□(□□ + □) = □□□
3×(42+2)=132

Digits: 8, 7, 7, 6, 1, 1, 1
(□ + □)(□ + □) = □□□
(7+6)×(8+1)=117

Digits: 8, 8, 8, 8, 2, 1, 0
(□ + □)(□ + □) = □□□
(8+8)×(8+0)=128

Digits: 9, 9, 9, 8, 8, 8, 2
(□ + □)(□ + □) = □□□
(9+8)×(8+9)=289

Digits: 6, 6, 6, 6, 4, 4, 1
(□ + □)(□ + □) = □□□
(6+6)×(6+6)=144

Digits: 9, 9, 9, 6, 5, 5, 1
(□ + □)(□ + □) = □□□
(9+5)×(5+9)=196

Digits: 9, 9, 2, 2, 2, 1, 1
(□ + □)(□ + □) = □□□
(2+9)×(9+2)=121

Digits: 9, 9, 4, 4, 4, 4, 1
(□ + □)(□ + □) = □□□
(9+9)×(4+4)=144

Digits: 8, 8, 6, 6, 2, 1, 1
(□ + □)(□ + □) = □□□
(1+8)×(6+8)=126

Digits: 8, 8, 6, 6, 6, 6, 1
(□ + □)(□ + □) = □□□
(6+6)×(6+8)=168

Directions: Place the given digits into the boxes to make the right-hand side as large as possible.

Digits: 1, 2, 3, 3, 9, 9
(□ + □□) × (□ + □□) =?
Answer: (1+93)×(2+93)=8930

Digits: 5, 5, 5, 6, 8, 9
(□ + □□) × (□ + □□) =?
Answer: (5+86)×(5+95)=9100

Digits: 2, 2, 3, 3, 8, 9
(□ + □□) × (□ + □□) =?
Answer: (2+92)×(3+83)=8084

Digits: 1, 3, 3, 5, 7, 8
(□ + □□) × (□ + □□) =?
Answer: (1+83)×(3+75)=6552

Digits: 4, 6, 7, 9, 9, 9
(□ + □□) × (□ + □□) =?
Answer: (4+99)×(6+97)=10609

Digits: 1, 6, 6, 6, 7, 8
(□ + □□) × (□ + □□) =?
Answer: (1+86)×(6+76)=7134

Digits: 1, 2, 5, 6, 8, 9
(□ + □□) × (□ + □□) =?
Answer: (1+92)×(5+86)=8463

Digits: 2, 2, 6, 6, 7, 8
(□ + □□) × (□ + □□) =?
Answer: (2+82)×(6+76)=6888

Digits: 2, 5, 6, 8, 8, 9
$(\Box + \Box\Box) \times (\Box + \Box\Box) = ?$
Answer: $(2+95)\times(6+88)=9118$

Digits: 1, 3, 3, 9
$\Box \times (\Box + \Box\Box) = ?$
Answer: $9\times(1+33)=306$

Digits: 2, 6, 9, 9
$\Box \times (\Box + \Box\Box) = ?$
Answer: $9\times(2+96)=882$

Digits: 5, 5, 5, 8
$\Box \times (\Box + \Box\Box) = ?$
Answer: $8\times(5+55)=480$

Digits: 2, 2, 3, 3
$\Box \times (\Box + \Box\Box) = ?$
Answer: $3\times(2+32)=102$

Digits: 1, 7, 8, 9
$\Box \times (\Box + \Box\Box) = ?$
Answer: $9\times(1+87)=792$

Digits: 3, 3, 5, 8
$\Box \times (\Box + \Box\Box) = ?$
Answer: $8\times(3+53)=448$

Digits: 4, 9, 9, 9
$\Box \times (\Box + \Box\Box) = ?$
Answer: $9\times(4+99)=927$

Digits: 6, 6, 7, 8
$\Box \times (\Box + \Box\Box) = ?$
Answer: $8\times(6+76)=656$

Digits: 1, 6, 6, 7
$\Box \times (\Box + \Box\Box) = ?$
Answer: $7\times(1+66)=469$

Digits: 1, 6, 8, 9
$\Box \times (\Box + \Box\Box) = ?$
Answer: $9\times(1+86)=783$

Digits: 1, 3, 3, 9
$(\Box + \Box) \times (\Box + \Box) = ?$
Answer: $(1+9)\times(3+3)=60$

Digits: 2, 6, 9, 9
$(\Box + \Box) \times (\Box + \Box) = ?$
Answer: $(2+9)\times(6+9)=165$

Digits: 5, 5, 5, 8
$(\Box + \Box) \times (\Box + \Box) = ?$
Answer: $(5+5)\times(5+8)=130$

Digits: 2, 2, 3, 3
$(\Box + \Box) \times (\Box + \Box) = ?$
Answer: $(2+3)\times(2+3)=25$

Digits: 1, 7, 8, 9
$(\Box + \Box) \times (\Box + \Box) = ?$
Answer: $(1+9)\times(7+8)=150$

Digits: 3, 3, 5, 8
$(\Box + \Box) \times (\Box + \Box) = ?$
Answer: $(3+5)\times(3+8)=88$

Digits: 4, 9, 9, 9
$(\Box + \Box) \times (\Box + \Box) = ?$
Answer: $(4+9)\times(9+9)=234$

Digits: 6, 6, 7, 8
$(\Box + \Box) \times (\Box + \Box) = ?$
Answer: $(6+7)\times(6+8)=182$

Digits: 1, 6, 6, 7
$(\Box + \Box) \times (\Box + \Box) = ?$
Answer: $(1+7)\times(6+6)=96$

Digits: 1, 6, 8, 9

$(\square + \square) \times (\square + \square) = ?$

Answer: $(1+9) \times (6+8) = 140$

CHAPTER 5

Diagrammatic Representations of Linear Systems

5.1   BOUQUET PUZZLES

A *bouquet puzzle* presents some bouquets of flowers, together with their price, and then presents another bouquet with an unknown price. These puzzles are intended to prepare students for solving simultaneous equations, known in some circles as systems of equations. An example of a bouquet puzzle appears in Figure 5.1.

What is the solution to the puzzle in Figure 5.1? If we compare the first and second bouquets, we see that the only difference is that a red flower was exchanged for a yellow one. This means that the red flowers cost $2 more than the yellow ones. Substitute a red flower for the yellow flower in the second bouquet. With three of the red flowers, the bouquet costs $15. This means that the red flowers cost $5 and so the yellow flowers cost $3. This tells us that the third bouquet costs $3 + $3 + $5 + $5 + $5 = $21. This solution method figures out the cost of each flower first, and then simply computes the price of the unknown bouquet, but there are sometimes easier methods.

The puzzle in Figure 5.1 had two types of flowers. You can have as many types of flowers as you want, but the problems get harder (on average) if there are more types of flowers. The puzzle in Figure 5.2 has three types of flowers. Let's solve it. Comparing the $12 dollar bouquet and the $10 dollar bouquet—which have two flowers in common—tells us that the green flower is $2 more expensive than the red one. If we take the $12 dollar bouquet and substitute a green flower for the red one, then its price goes up to $14 *and* it becomes the target bouquet. Pretty slick method, isn't it? Notice that the $13 bouquet is not used—it is something of a red herring.

There are usually several ways to solve a bouquet problem, making it good for a class exercise where students solve a problem in small groups and then compare their solutions. The question is asked in a way that does not suggest a solution technique and, better still, the easiest method is different for different problems. This means that bouquet puzzles are a nice type of problem for developing problem solving strategies and skills. It is also the case that, for some of these puzzles, we can solve the puzzle *without* having to find the individual price of each type of flower.

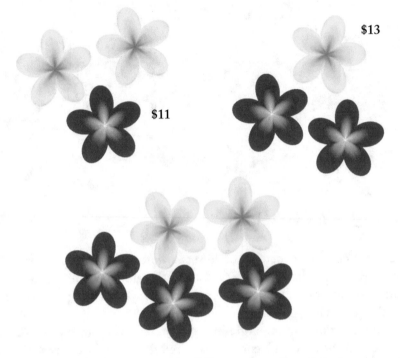

Figure 5.1: An example of a bouquet puzzle. What is the price of the bouquet with 2 yellow flowers and 3 red flowers?

5.1.1 THE CONNECTION OF BOUQUET PUZZLES TO VECTORS

Bouquet puzzles are actually very simple structures built out of vectors, with pictures added to finish the puzzle. If we look at the puzzle in Figure 5.1 then it consists of the vectors (2,1), (1,2), and (2,3). The first coordinate is the number of yellow flowers, the second, the number of red flowers. We also have a price vector (\$3, \$5). The prices of the bouquets are dot products of the vectors describing the bouquets with the price vector:

$$(2, 1) \cdot (\$3, \$5) = \$6 + \$5 = \$11$$
$$(1, 2) \cdot (\$3, \$5) = \$3 + \$10 = \$13$$
$$(2, 3) \cdot (\$3, \$5) = \$6 + \$15 = \$21.$$

A **dot product** is a way to multiply two vectors together and get a single value as a result. The first coordinates get multiplied together, and the second coordinates get multiplied together; then the results of those multiplications get added into a single number.

This is where the problem factory effect kicks in. First, you make the sample bouquets and choose a price vector. You then add and subtract the sample bouquets a few times, making sure to end with a vector with no negative entries, to get the bouquet with an unknown price. The

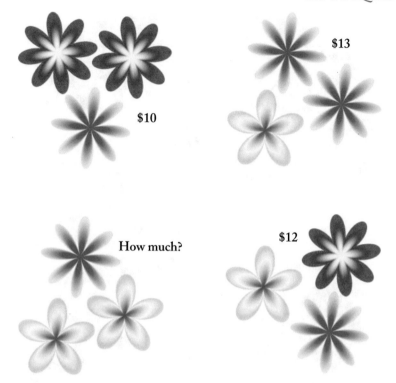

Figure 5.2: Another example of a bouquet puzzle.

price of any bouquet can be computed by taking the dot product of the vector associated with the bouquet with the price vector. For the problem in Figure 5.2 the vectors are, in (red, yellow, green) order, $10-(2, 1, 0), $13-(0, 2, 1), $12-(1, 1, 1), and (0, 1, 2). The price vector is ($3, $4, $5).

To reiterate a point using the new terminology we've just developed, it may not be possible to solve for the price vector from the data in a bouquet problem. When you are constructing the problem, you have *a* price vector; for some problems a second price vector might yield all the same prices. This does not change the answer to the problem and so it is only an issue because some students may want to solve the problem by figuring out the price of the individual flowers. When you encounter one of these problems, it can be a good place for a sermon on problem solving.

A variation on bouquet puzzles arises from noticing that there is no particular need to have only one bouquet with an unknown price. The puzzle shown in Figure 5.3 has three example bouquets with prices and two target bouquets. Here, information computed while finding the price of one of the targets may be useful for solving the other. For the third puzzle, if we subtract the $19 bouquet from the $27 bouquet we observe that two purple flowers minus two blue

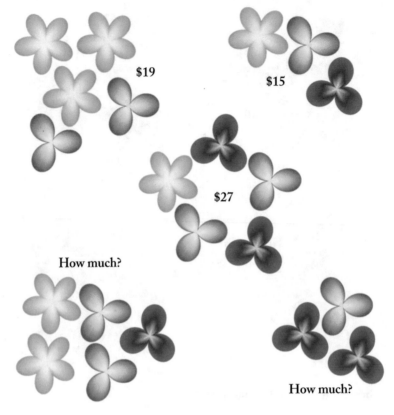

Figure 5.3: A bouquet puzzle with two bouquets with no price.

flowers is worth $8 and so, dividing by 2, that the purple flowers cost $4 more than the blue ones. If we use this fact on the $15 bouquet, substituting a purple for a blue, we get that purple, purple, yellow costs $19—and this is one of the targets!

If we take the $27 bouquet, add in the $15 bouquet, and then subtract the bouquet that we figured out costs $19 then we have two blue, two yellow, and one purple flower. This is the other target bouquet and we see that it costs $27 + $15 − $19, or $23 and we have solved for the price of both target bouquets. Notice that this solution requires matching patterns in the available bouquets, another important skill.

If you have a large number of bouquets of unknown price then you can treat the problem as a speed run: figure out the missing prices as fast as possible. For example, "A flower shop owner comes in to find that many of the labels for his ready-to-go bouquets of flowers have been modified or vandalized. These bouquets are shown below. The owner was able to remember a few of the prices and, while he talks to the police, wants you to figure out the others and make new signs. He wants this done before the shop re-opens so please find the correct prices as soon as

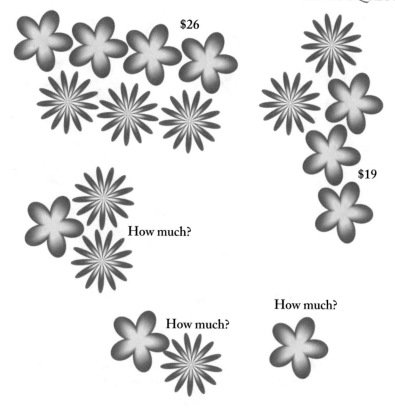

Figure 5.4: A bouquet puzzle with three bouquets with no price.

possible!" A small problem of this sort appears in Figure 5.4. Notice that the owner remembered the price of the two most expensive items.

Solving the problems in Figure 5.4, subtracting the $19 bouquet from the $26 bouquet immediately tells us the two-flower bouquet has a price of $7. Subtracting three copies of the two-flower bouquet from the most expensive bouquet gives us the result that the blue flower costs $5 and so the violet flower costs $2. This then makes it easy to see that the three-flower bouquet costs $9. Notice that the price vector can be deduced in this problem and that re-use of earlier results leads to a very efficient solution path.

Something mentioned in the text is that a bouquet puzzle that can be solved may not permit determination of the prices of individual flowers. Figure 5.5 shows an example of such a puzzle. There are a group of five identical flowers in each example and the solution. This means there is no way to tell the price of the two flowers making up that group of five. This is a feature of the puzzle, not a bug, and it may be useful to deterring students that insist on solving for the price of individual flowers.

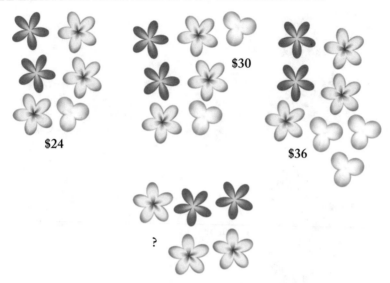

Figure 5.5: A bouquet puzzle were it is impossible to determine the individual price of two of the flowers.

A more extreme version of this can be achieved by having fewer example bouquets than there are types of flowers. If we look at the underlying linear algebra, this means that solving for individual prices is not possible because the corresponding system of equations does not contain enough information. In Section 3.2 we already assumed that students can solve three equations in three unknowns—bouquet puzzles prepare students for this activity, but they also can often be solved my much easier methods than breaking out the simultaneous equation toolkit. If you have a student that has mastered simultaneous equations, handing them an under-specified problem may create a need to think things through, rather than just plug-and-chug.

5.1.2 MAKING YOUR OWN PUZZLES

Let's begin with a quick review of the dot product, which is used in creating bouquet problems. If you have two vectors $\vec{v} = (v_1, v_2, \ldots, v_k)$ and $\vec{u} = (u_1, u_2, \ldots, u_k)$ then

$$\vec{v} \cdot \vec{u} = v_1 \cdot u_1 + v_2 \cdot u_2 + \cdots + v_k \cdot u_k.$$

You just multiply the corresponding coordinates and add the results.

If you just make up bouquets and prices, the chances are good that you will get a problem that cannot be solved or one that involves a lot of fractions. The key is to use the vector mechanics outlined earlier in the chapter. Choose the example bouquet vectors: $B_1 = (2, 2, 2)$, $B_2 = (1, 2, 3)$, and $B_3 = (3, 2, 1)$ and a price vector ($3, $5, $6). Now create a target bouquet

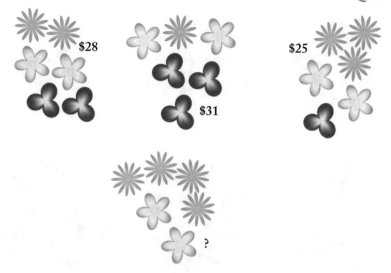

Figure 5.6: Target bouquet with one of the flower types dropped out.

but doing simple arithmetic on the example vectors

$$B_t = B_1 - B_2 + B_3 = (4, 2, 0).$$

This target B_t has no negative coordinates but it does have a zero coordinate which makes the problem more interesting. Since we are working with three-coordinate vectors there are three sorts of flowers. Taking the dot product with the price vector we get that the costs are $28, $31, $25 for the examples $B_1 - B_3$ and the target bouquet has a price of $(4, 2, 0) \cdot (\$3, \$5, \$6) = \$22$. Now all we need to do is make the picture (see Figure 5.6).

This problem is the second one to have one of the flower types drop out in the target bouquet. This does not mean you do not need the information conveyed by the missing flower. Here are a few tips for making problems.

- Changing the price vector and re-computing all the prices gives you a fast way to generate new puzzles—in particular the picture you already have will still do for the new puzzle, except for the numbers.

- If you have a puzzle that permits determining the value of individual flowers then asking the students to find the per-flower cost of each type of flower is not a bad problem.

- Increasing the number of types of flowers increases the dimension of the problem and, usually, its difficulty.

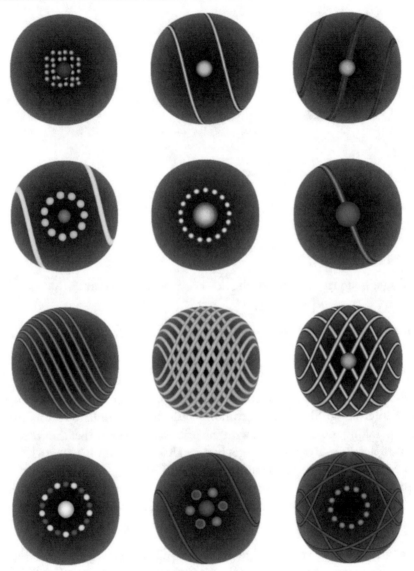

Figure 5.7: Chocolate truffles.

5.2   RESKINNING BOUQUET PUZZLES

Flowers are a convenient way to picture diagrammatic linear systems, but there are many other possible ways to picture these problems, called a "skin." Objects that would work as skins for these puzzles should naturally come in groups and have variable prices. The second skin for our problems is chocolate truffles (Figure 5.7). The first of our diagrammatic linear systems with

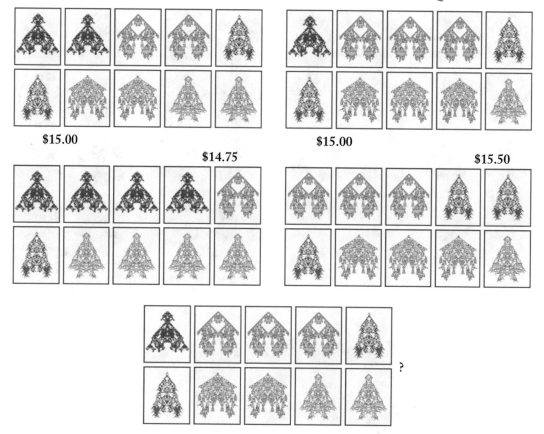

$15.00

$15.00

$14.75

$15.50

?

Figure 5.8: Four hands of cards with known prices and one unknown hand of cards. The unknown price is $15.25.

truffles is shown in Figure 5.9. The second is shown in Figure 5.10. At the beginning of the section are 12 images of chocolate truffles that you can use for making your own problems. You can download the images from the publisher's website.

Another natural candidate for a skin for these problems are collectible cards. If there is a type of collectible card that your students are interested in, then those could easily be used to pose problems (and perhaps used as prizes). We have also made up some collectible cards from mathematical pictures which can be used as well. If you are using some other sort of card, even playing cards, you can do one-to-one substitution of cards into the problems we pose.

Our first problem posed with trading cards appears in Figure 5.8. This problem has more types of items that earlier problems as well as very close prices and a duplicate price. This may concern some students, but the problem can still be solved. This example is not only an additional skin, it is an escalation of the difficulty.

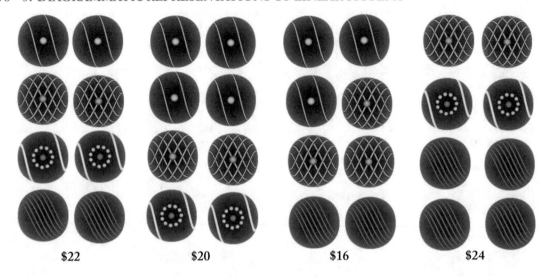

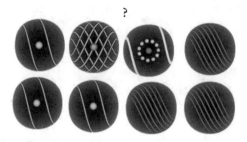

Figure 5.9: Five boxes of eight truffles, one with an unknown price. The unknown price is $17.

The second trading card problem uses a variety of small (two-card) examples and poses three targets, also with two cards. The examples are rich in patterns that can help students to come up with a solution method. The problem is posed in Figure 5.11. These examples correspond to vectors $(1, 1, 0, 0, 0)$, $(0, 1, 1, 0, 0)$, $(0, 0, 1, 1, 0)$, $(0, 0, 0, 1, 1)$, and $(1, 0, 0, 0, 1)$. The price vector for this problem is ($0.75, $1.25, $1.75, $1.50, $1.00), which makes it easy for you to construct additional problems from the 5 two-card hands in Figure 5.11.

5.2.1 SOLUTION TECHNIQUES

A practical problem for these problems is trial addition and subtraction. Let's solve the problems in Figure 5.11. If we add all five examples we get that two of each card costs $12.50 and so one of each card costs $6.25. Subtracting pairs of examples that leave one card gives us that the cost of the red-top-blue-bottom card is $0.75, the green card with more internal pattern is $1.00,

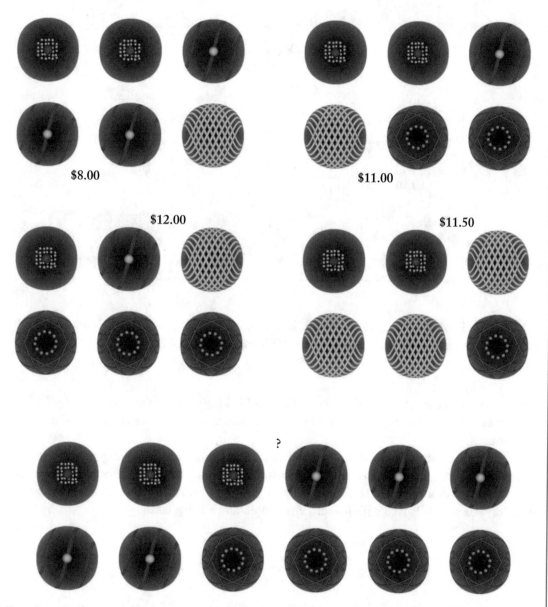

Figure 5.10: Four boxes of 6 truffles with known prices and one with 12 that has an unknown price. The unknown price is $19.50.

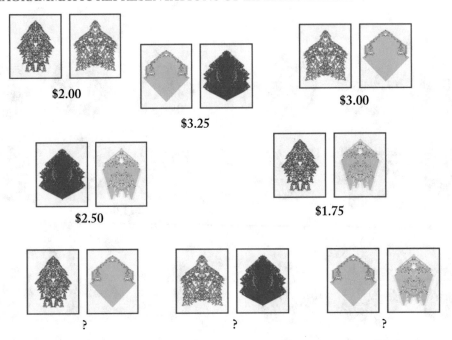

Figure 5.11: Four hands of two cards with known prices and 3 two-card hands with unknown prices. The unknown prices, left to right, are $2.50, $2.75, and $2.75.

the mostly violet card is $1.25, the mostly blue card is $1.50, and the mostly green card $1.75. This gives us the values of each card, and we see the answers given in the caption of Figure 5.11 are correct.

Encourage students to develop an abstract notation, since sketching pictures for the intermediate steps is cumbersome. In the abstract notation, adding all five examples yields $(2, 2, 2, 2, 2)$ with a price of $12.50, for example, and so $(1, 1, 1, 1, 1)$ costs $6.25. The abstract notation works equally well with flowers, chocolates, and trading cards.

5.3 ASSEMBLING PACKAGES WITH A GIVEN VALUE

This section deals with a different use for the props we have created for bouquet puzzles and their re-skinned cousins. A practical problem is, given a collection of goods with particular, individual prices, assemble a package of goods with a particular value. The authors think that this activity is most natural for the chocolates, among the props we have established thus far, and so we being with our chocolates as a price list, given in Figure 5.12.

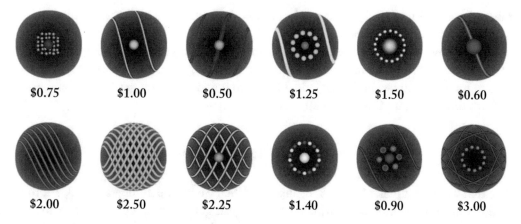

Figure 5.12: A price list for individual chocolates.

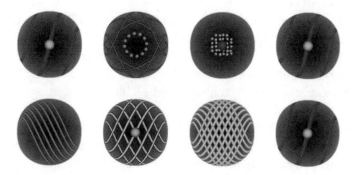

Figure 5.13: Box of chocolates.

Example 5.1 This example poses a package assembly problem—but one where a good part of the package is already specified. This is a way of taking something that looks like a large problem and turns it into a smaller problem.

A customer asks for a gift box with one each of the four most expensive chocolates, eight chocolates total, and costing $12.00. Using the price list, we total up the candies that must go in the box and get $9.75 leaving $2.25. We need to lean heavily on the $0.50 filler candies but the problem can be solved with the box of chocolates shown in Figure 5.13.

5.3.1 MOST ASSEMBLY PROBLEMS HAVE SOLUTIONS

There is some interesting math underneath the package assembly problem. The short version of this is that if a problem can be solved at all, it can usually be solved several ways or, if the number of items in the package is large, in millions of ways. Let's look at a simple instance of this problem with only two prices.

Suppose we had the numbers 3 and 7 (our prices) and wanted to know which numbers we could make by adding them, with an unlimited supply of 3's and 7's. Start by listing 3 and 7 and then, for each number in the list, add 3 and 7 to it and see if you get a new member of the list. If we continue up until we reach 20 in the list we get this:

$$3\,6\,7\,9\,10\,12\,13\,14\,15\,16\,17\,18\,19\,20.$$

Notice that we cannot make 1, 2, 4, 5, 8, and 11. When we hit the numbers 12, 13, 14, then we have three consecutive numbers—which means we can get anything by continuing to add 3's. To put it another way, you can get any large whole number different from one of 12, 13, or 14 by adding a positive multiple of three to one of those numbers. We deduce that all but six of the whole numbers are a sum of 3's and 7's.

This phenomenon is general, with one limitation that we will explain in a minute. When you have a small set of numbers (prices of chocolates), then there are are some numbers (prices) that you cannot hit with packages that are possible to assemble. After that you can hit every possible price with at least one assortment of chocolates. This means that, while working out the details of a chocolate assortment problem may be a bit tricky, finding assortments that are possible to make is pretty easy.

The limitation is this: if the numbers all have a common divisor, then every number you can make must be a multiple of that common divisor. If every number in your original set of numbers is even, then you can only reach an even number, for example. In the price schedule for the chocolates at the beginning of this section, the common divisor is 5 cents, so every price will be a multiple of 5 cents. Given that limitation, we can run the "keep adding numbers" algorithm, on our computer, to find all the prices that are impossible to build up with any number of chocolates. Anything below 50 cents, of course, as well as 55 cents, 65 cents, 70 cents, 80 cents, 85 cents, 95 cents, $1.05, $1.15, $1.30, $1.45, $1.55, and $2.05. All other prices are possible.

The example also wanted not just a price, but also an 8-piece box, which is an additional constraint, as well as specifying several things that had to go in the box. The problem was still possible to solve, but there was not a lot of maneuvering room. In general, the fact that most of these problems are possible makes it easier for an instructor to design problems.

Before we move to more examples, let's look at one more factor. Suppose that we fill a box that costs $12.00 by just putting in chocolates, one at a time, until it is full. Then, including the order in which we put in the chocolates, there are a little over 3.75 billion ways to do this! This is not the number of different possible selections of chocolates; there are only a few thousand of those. Rather, it is the number of ways someone filling a $12.00 box at random could get there. This number is only useful because it again emphasizes that these problems have a lot of room when you are solving them.

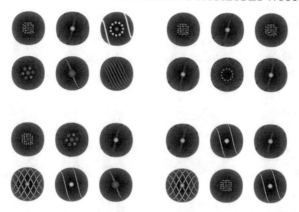

Figure 5.14: Starter set of four chocolate boxes.

Example 5.2   We want to sell $6.00 boxes of six chocolates—and we want to have a number of choices. Find as many boxes of six chocolates that cost $6.00 as you can.
Shown in Figure 5.14 is a starter set of four such boxes. There are many others.

Suppose we order problems by the number of chocolates in the solution. As we increase the number of chocolates, the number of solutions to a problem that has a solution at all tends to explode. This is a mathematical phenomenon called a *combinatorial explosion*. We will see this phenomenon again in Chapter 6 where the number of different ways to assemble puzzle pieces into a rectangle or square will turn out to be enormous.

Combinatorial explosions are often associated with problem factories. In the popular puzzle Sudoku, some of the entries in the puzzle are filled in. When the customer in our first package assembly problem required the four most expensive chocolates be part of the package, they were filling in some of the pieces of the puzzle. If there are a combinatorial explosion's worth of examples, then we can posed the $6.00 box problem with partial solutions secure in having many, many possible solutions to work toward.

Example 5.3   A larger example would be to find a box of twelve chocolates that costs $30.00. One possible solution is shown in Figure 5.15.

Here are some options for posing these problems. Remember to check that there is a solution unless your goal is to ask the question "is this possible?"

- Limit the number of chocolates in the box.

- Give a price for the box of chocolates. Slightly easier, put a maximum price on the box.

- Specify some of the chocolates in the box to start with.

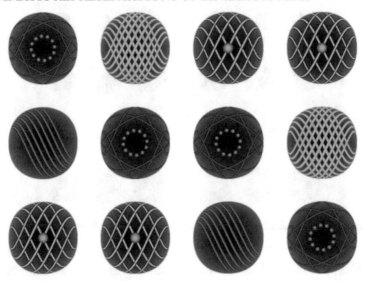

Figure 5.15: A box of 12 chocolates that costs $30.00.

- Demand no repeats or, alternatively, that there be two or three of each chocolate used.

A possible competition is to have a class design as many boxes of a given size and price as possible in a fixed amount of time. The four $6 boxes of six chocolates can serve as an example of this goal.

5.3.2   OTHER TYPES OF QUESTIONS YOU CAN ASK

The problem of making a box of chocolates is a pretty good problem factory. It permits you to pose hundreds of problems small enough that students can do them. Problems are very likely to have a solution, or many solutions, making them easier to solve. These problems are suitable for many grade levels—use more constraints or larger boxes for higher grades. These puzzles are also related to using vectors. A vector is a list of numbers specifying distances (or amounts) in different directions, like "three kilometers north and five kilometers east." Vectors are the basis of a lot of advanced math. When we are making boxes of chocolates, the types of chocolates correspond to different directions and the number of chocolates of a particular type tell you how far you go in that direction. This makes chocolate box problems a way of getting students ready for vectors. This said, there are some pure math problems that come up in figuring out the math underlying the chocolate box problem which might also make good questions for students.

Given a small set of numbers, like 3 and 7 in our example above, the following questions are potentially interesting and challenging for lower level students.

1. Given the numbers, is there a common multiple that limits which numbers we can make by adding them? The remainder of the questions assume there is no such common divisor.

2. List the numbers you cannot make by adding these numbers.

3. Find the largest (or last) number you cannot make by adding these numbers.

4. Find the smallest number that can be made two different ways by adding these numbers.

The rule for figuring out that you can now make any number after a given point is to determine if that point is the beginning of a run of numbers as long as the smallest number you are allowed to add. For the example with 3 and 7 we could not make 11, but we could make 12, 13, and 14—a run of three numbers. Any larger number is a multiple of 3 plus one of 12, 13, or 14, so after 11, we can make anything.

5.4   EXAMPLE PUZZLES AND PROBLEMS

The images of flowers, trading cards, and chocolates may be obtained from the publisher's website to let you assemble your own problems. There are also a number of problems in the text of the chapter that you can use. Here we will give some additional problems and solutions, but many of them are in a form without a skin.

5.4.1   BOUQUET PUZZLES

Here are four groups of five bouquet puzzles, ordered by increasing complexity.

Examples: (1,1,2) $19; (1,2,0) $12;
Target: (3,4,4) $50

Examples: (0,1,2) $11; (2,2,1) $15;
Target: (4,3,0) $19

Examples: (1,0,1) $6; (2,2,2) $14;
Target: (3,4,3) $22

Examples: (1,2,1) $12; (2,2,0) $8;
Target: (3,4,1) $20

Examples: (1,0,1) $8; (1,2,2) $12;
Target: (2,0,2) $16

Examples: (1,2,0) $12; (2,2,1) $23; (2,1,0) $15;
Target: (3,3,0) $27

Examples: (2,1,1) $14; (0,1,1) $8; (0,1,2) $10;
Target: (4,3,4) $38

Examples: (2,1,0) $13; (0,2,1) $12; (1,2,1) $16;
Target: (2,3,1) $25

Examples: (0,1,1) $7; (2,2,0) $6; (1,2,2) $15;
Target: (1,2,0) $5

Examples: (2,2,1) $12; (1,1,1) $8; (0,2,1) $6;
Target: (2,6,4) $28

Examples: (2,1,2,0) $25; (0,1,0,2) $5; (1,1,1,2) $16;
Target: (3,4,3,6) $51

Examples: (2,2,2,1) $25; (2,1,0,1) $14; (1,1,1,2) $14;
Target: (3,2,1,0) $25

Examples: (0,2,0,1) $17; (1,2,0,0) $15; (1,2,2,2) $29;
Target: (1,2,4,4) $43

Examples: (1,0,0,1) $7; (0,0,2,1) $7; (1,2,1,2) $22;
Target: (2,2,1,3) $29

Examples: (0,0,2,1) $7; (1,2,0,0) $10; (2,2,2,2) $26;
Target: (2,2,6,4) $40

Examples: (2,1,2,0) $25; (0,1,0,2) $5; (1,1,1,2) $16; (2,0,0,1) $13;
Target: (3,2,3,0) $39

Examples: (1,1,2,0) $16; (0,1,2,2) $20; (2,1,1,1) $16; (1,0,1,0) $7;
Target: (3,3,4,1) $41

Examples: (1,1,1,0) $10; (1,0,1,1) $6; (2,0,1,1) $9; (0,1,1,2) $9;
Target: (2,2,4,6) $30

Examples: (1,2,0,0) $12; (1,0,2,1) $19; (2,0,0,2) $22; (1,1,1,2) $23;
Target: (3,2,0,2) $34

Examples: (2,0,2,0) $16; (2,1,1,0) $18; (0,2,0,1) $9; (0,1,1,1) $7;
Target: (2,1,3,1) $23

5.4.2   CANDY BOX PROBLEMS

Here are some alternate lists of prices for candies, cards, or flowers and the prices that they cannot achieve for making your own packaging problems. Any price not on the list of impossible prices is possible—but if it is large relative to the prices of the good it will require a large number of items and will be relatively easy. The ease of solution arises from the fact there are many ways to assemble the package.

Prices for six types of items: $0.25, $0.35, $0.50 $0.65, $0.75, and $1.00. An assortment of objects with these prices can have an price that is a multiple of 5 cents, except prices below $0.25 and the prices $0.30, $0.40, $0.45, $0.55, and $0.80.

Prices for eight types of items: $1.75, $3.00, $3.50, $4.25, $5.00, $6.00, $7.00, and $7.50. An assortment of objects with these prices can have any price that is a multiple of a quarter, except prices below $1.75 and the prices $2.00, $2.25, $2.50, $2.75, $3.25, $3.75, $4.00, $4.50, $5.50, $5.75, and $6.25.

Prices for six types of items: $8, $12, $15, $17, $23, and $26. An assortment of objects with these prices can have any price that is a multiple of a dollar, except prices below $8 and the prices $9, $10, $11, $13, $14, $18, $19, $21, and $22.

Prices for nine types of items: $0.04, $0.07, $0.11, $0.15, $0.19, $0.23, $0.25, $0.29, $0.31. An assortment of objects with these prices can have any price, except prices below $0.04 and the prices $0.05, $0.06, $0.09, $0.1, $0.13, and $0.17.

Prices for seven types of items: $0.60, $0.70, $0.85, $0.95, $1.10, $1.25, and $1.35. An assortment of objects with these prices can have any price, except prices below $0.60 and the prices $0.65, $0.75, $0.8, $0.9, $1, $1.05, $1.15, $1.5, $1.6, and $1.75.

Prices for six types of items: $3.25, $4, $4.75, $5.5, $6, and $6.75. An assortment of objects with these prices can have any price, except prices below $3.25 and the prices $3.5, $3.75, $4.25, $4.5, $5, $5.25, $5.75, $6.25, $7, $7.5, $7.75, $8.25, $8.5, $9, and $11.75.

Prices for five types of items: $0.04, $0.07, $0.12, $0.2, and $0.22. An assortment of objects with these prices can have any price, except prices below $0.04 and the prices $0.05, $0.06, $0.09, $0.1, $0.13, and $0.17.

Prices for six types of items: $1.00, $1.75, $3.00, $4.25, $5.00, and $5.50. An assortment of objects with these prices can have any price, except prices below $1.00 and the prices $1.25, $1.5, $2.25, $2.5, and $3.25.

CHAPTER 6

Polyomino Tiling Puzzles

A *polyomino* is a shape made of squares, all the same size, joined to one another along full faces. Figure 6.1 shows the polyominos with four or fewer squares. The number of squares in a polyomino is called the polyomino's *size*.

The different size classes of dominoes are named for their size. The single square by itself is a **monomino**. The two-square polyomino is called a **domino**. The three- and four-square polyominos are called **tromioes** and **quadrominoes**, respectively. An important point to make is that the polyominos in Figure 6.1 are shown on one orientation each. In fact, by rotating the polyomino or flipping it over, there are as many as eight different forms of a polyomino, though some, like the 2×2 square quadromino, do not change when you rotate or flip them.

6.1 INDECOMPOSABLE TILINGS WITH DOMINOES

Our first problem factory is posing problems about filling rectangles with dominoes. This is one of the simplest sorts of problems you can pose, and simply changing the size of the rectangle being filled gives you new problems. Filling a rectangle with a given polyomino is called *tiling the rectangle* with that polyomino. We're going to start off with a simple, obvious piece of theory that is a little useful for tilings with dominoes and very useful later in the chapter.

Fact: If a rectangle can be tiled with a polyomino of size n then the area of the rectangle must be a multiple of n. To see this, the number of squares in the rectangle can be divided into groups, covered by k different copies of the polyomino, that are a witness that the area of the rectangle is kn.

The application of this fact to tiling with dominoes is that a rectangle can only be tiled with dominoes if its area is even. The question "can this rectangle be tiled with dominoes?" is not that interesting without some additional conditions. The answer is "yes, if it has even area." To that end, let's add some additional conditions.

A natural condition is to specify the number of vertical and horizontal dominoes used with the simplest goal being to make them even. An example of such a tiling is shown in Figure 6.2. Notice that if the rectangle is $4n \times 2m$ that a generalization of this solution can be used on any such rectangle.

The cases of all horizontal and all vertical are possible if the width (for horizontal dominoes) or the height (for vertical dominoes) are an even length. Another interesting case is to ask the students to use an odd number of vertical or an odd number of horizontal dominoes. Finally,

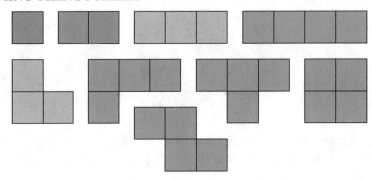

Figure 6.1: The polyominos of size at most four.

Figure 6.2: A tiling of a 4 × 4 rectangle with equal numbers of vertical and horizontal dominoes.

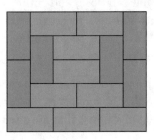

Figure 6.3: A tiling of a 5 × 6 rectangle with six vertical and nine horizontal dominoes.

simply specifying a target number of vertical or horizontal dominoes, scoring students on how close they get. The problem of filling a rectangle with dominoes, with whatever constraint, can be posed as a problem or presented as a competitive activity.

In the spirit of problem factories, making sure that you have a solvable problem is best done by tiling a rectangle and counting the vertical and horizontal dominoes. An example appears in Figure 6.3 with six vertical and nine horizontal tiles.

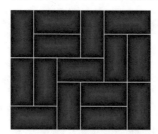

Figure 6.4: An example of an indecomposable tiling with dominoes.

6.1.1   THE INDECOMPOSABILITY CONSTRAINT, A LARGE EXAMPLE

If a tiling of a rectangle cannot be split, vertically or horizontally, into two rectangles, we call it *indecomposable*. If it can be split then we call it *decomposable* and the split is a *decomposition*. An example of of an indecomposable tiling appears in Figure 6.4.

Indecomposability is an additional condition that can be used to pose domino tiling problems, but there is an inobvious constraint on using indecomposability as a condition. It turns out that the example in Figure 6.4 is the smallest possible indecomposable tiling with dominoes.

Claim 6.1   No rectangle can be tiled indecomposably with dominoes unless it's smallest dimension is at least five, with the exception of a 2×1 rectangle filled with a single tile.

Proof: We look at dominoes with a dimension less than five in order of their smallest dimension. Consider an arrangement of bricks that is only one high (see Figure 6.5a).

If there is more than one brick then it is trivial to find a vertical line that splits the arrangement into two rectangles. Now consider configurations of height two (Figure 6.5b).

If the first brick is vertical, there is a vertical line that splits the arrangement into two rectangles. If we don't start with a vertical brick then we have to start with two horizontal bricks. These, however, form a square—a type of rectangle—and again a vertical line can divide the arrangement into two rectangles. So, with the exception of "one brick" there are no one-high or two-high indecomposable configurations.

Consider three-high configurations like the one in Figure 6.5c.

If we start with three horizontal bricks, stacked vertically, we have a decomposable configuration. If we start with three horizontal bricks and continue, we can split off the initial 2×3 rectangle. From this we deduce there must be one vertical brick in the first column. This forces a horizontal brick spanning the first two columns (this is shown above). If there is a vertical brick in the second column it completes a rectangle that can be split off—and that rectangle can itself be split. This forces the two horizontal rectangles spanning the second and third column but that forces the second horizontal brick in the bottom row. This situation, two horizontal, one horizontal, and so on, continues until we end with a vertical brick. This, however, causes

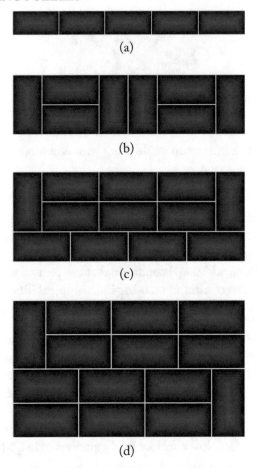

Figure 6.5: (a) Arrangement of bricks that is only one high; (b) configurations of height two; (c) three-high configurations; and (d) four-high configurations.

the entire last row to be a rectangle that can be split off. If we placed the first vertical brick differently, a variation of the above situation would still yield a horizontal split, just a different one. We conclude that no indecomposable arrangement of bricks can be three-high. Now we turn to the four-high configurations. The one shown in Figure 6.5d is exemplary.

Following the same reasoning as held in the three-high case, there must be both vertical and horizontal dominoes in the first column of the tiling. There are three ways to arrange these, with the vertical tile at the top, in the middle, or at the bottom of the column, and all three possibilities yield a decomposable tiling if we fill in the vertical domino that ends at 4×2. If, on the other hand, we add two more horizontal rectangles (the only other possibility) we achieve a situation where, again, completing the tiling with a vertical domino yields a decomposable configuration.

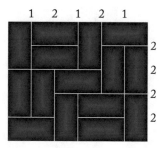

Figure 6.6: The split prevention numbers along each potential dividing line of our minimal 5 × 6 indecomposable tiling.

This situation, where we have the choice of ending with a vertical tile or continuing with two horizontal ones, is persistent—if the rectangle has finite length it ends being possible to split it along the middle horizontally or peeling off the top or bottom row, depending on the placement of the original vertical tile. We conclude that there are no indecomposable tilings of height one through four. □

Notice that we gain four questions with increasing levels of challenge from the four cases of the tilings that cannot be indecomposable, with shortest dimensions 1–4. The real value of this claim is to avoid asking students to try to locate indecomposable rectangles that do not exist. To that end we add one additional result.

The picture shown in Figure 6.6 is the 6 × 5 indecomposable arrangement, but this time it has been annotated with the number of bricks that prevent the arrangement from having a decomposition. These numbers are in line with each of the potential vertical and horizontal lines where a split might be possible. If you compute these numbers for a configuration that has a decomposition, then one of them is zero. This means that computing these numbers, which we call *split prevention numbers*, gives us an algorithm for checking indecomposability of a tiling. Notice that, since every brick blocks one possible split, the split prevention numbers add up to the number of bricks in the tiling.

Claim 6.2 There is no indecomposable domino tiling of a 6 × 6 configuration.

Proof: Consider a 6 × 6 tiling. Since the space taken up by vertical bricks in the first column must be even, the number of horizontal bricks starting in the first column and going right must also be even. This means the split prevention number between the first two columns is even. The same reasoning, applied to the space remaining in the second column, forces the split prevention number between the second and third column to be even. This continues, and the same logic can be applied to pairs of rows. We can conclude that, for a 6 × 6 arrangement, all the split prevention numbers must be even.

The smallest even number bigger than zero is two. Remembering that if there were any zeros then the configuration could be decomposed, we deduce that the split numbers are all *at*

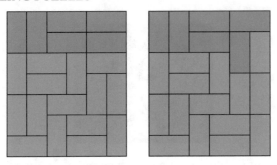

Figure 6.7: Extending an indecomposable tiling.

least 2. There are five pairs of rows and five pairs of columns in a 6 × 6 arrangement where we need to prevent a split. That means that the numbers must be (at least) 2, 2, 2, 2, 2, 2, 2, 2, 2, 2: ten twos. This, in turn, means there are 20 bricks, each taking up 2 squares. That is a total of 40 squares that the bricks cover, but a 6 × 6 arrangement contains only 36 squares. We conclude that a 6 × 6 configuration cannot be indecomposable. □

Sizes of rectangles with both dimensions exceeding five that are also not 6 × 6 (and even) are safe to assign for locating indecomposable tilings.

6.1.2   SOLUTION TECHNIQUES

When we are trying to solve the problem of tiling a rectangle with dominoes, using a particular number of vertical and horizontal dominoes, then there are some general techniques. Break up the rectangle into sub-rectangles and use the last one of balance things. Also, if you have two adjacent dominoes that form a 2 × 2 square, they can be rotated a quarter-turn to shift the balance of vertical and horizontal by two. Usually, these puzzles are not too difficult. The technique of rotating a 2 × 2 square, however, is also useful for making indecomposable tilings.

Figure 6.7 shows how to take the original 5 × 6 indecomposable tiling and extend it to a 7 × 6 indecomposable tiling. The process starts by extending the tiling, shown in blue, with a 2 × 6 block, shown in red. This, of course, yields a decomposable tiling. Rotating a well-chosen 2 × 2 block restores the indecomposability of the tiling.

You may want to demonstrate this technique to your students *after* asking them to try to find larger indecomposable tilings. The example in Figure 6.7 extended by two in the vertical direction to avoid the impossible 6 × 6 configuration, but in other cases, when the length of the block is even, you can add a single layer of blocks and still rotate a 2 × 2 square. If you try this with the 5 × 6 example, it fails, which might be an interesting point to explore.

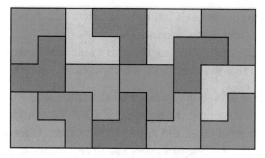

Figure 6.8: An example of tiling a 5 × 9 rectangle with the L-shaped tromino.

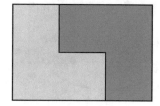

Figure 6.9: The smallest L-tromino tiling.

6.2 WHEN CAN YOU TILE A RECTANGLE WITH A POLYOMINO?

Our next problem factory looks at a type of puzzle where you want to fill a rectangle with a given polyomino. We will be using the L-shaped 3-square polyomino, used to fill a 5 × 9 rectangle in Figure 6.8, as our example to motivate the problem. The goal is to figure out every possible size of rectangle that can be filled with this shape. In the last section we did this for dominoes, and the answer was any rectangle with even area. If you are constructing puzzles for other people—e.g., your students—knowing which problems can be solved gives you an edge. The problem factory has two branches—asking if a given rectangle can be filled with a specific polyomino and, this is much harder, asking what all the sizes of rectangles are that can be filled with a polyomino.

If we are tiling rectangles with the L-tromino then our first step is to use the fact which tells us that the rectangle must have an area that is a multiple of three, because it is covered with objects of area three each. This is a step you take for every problem asking if a given polyomino tiles a rectangle. In particular, when a rectangle has an area that is *not* a multiple of the area of the polyomino, then provides a quick proof that the tiling problem is impossible; frequently students will burn a lot of effort on trying to find an impossible tiling before trying this simple "is it possible?" test.

Figure 6.9 shows the smallest possible rectangle that can be tiled by an L-tromino. This 3 × 2 or 2 × 3 rectangle can be used to tile any rectangle that has a side of even length with the other size having a length that is a multiple of three, which buys us an infinite family of rectangles that can be tiled, and $3n \times 2k$ rectangle.

This leaves us with the case of rectangles with a side length that is a multiple of three by one that is of odd length. Here we begin with an impossibility result.

Fact: A rectangle with one side length three and the other odd cannot be tiled with an L-tromino. To see this, we start by demonstrating that a 3×3 rectangle cannot be tiled with the L-tromino. An L-tromino must cover the upper-right corner. There are three ways to do this, but by rotating or flipping the 3×3 square, they are all equivalent to the situation shown in the picture below which generates an unfillable 3×1 remainder.

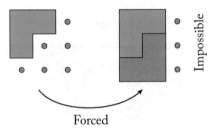

Forced

Now consider a rectangle that is three high by any odd number. Every way of placing the shape in the upper left corner does one of three things illustrated below: (i) do something impossible right away, (ii) shorten the rectangle into a 3×2 shorter rectangle (this can happen two ways), or (iii) create a 3×1 space that cannot be filled. The picture shows all three cases for a 5×3 rectangle.

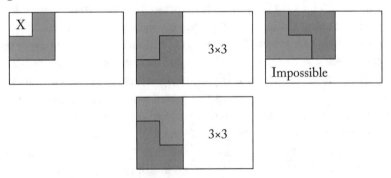

Of these possibilities, the only viable one fills the right-most two columns of the rectangle, leaving a $3 \times (n - 2)$ space. Filling that space shortens the rectangle two more, over and over, until we reduce the problem to the impossible 3×3 case. We deduce that $3 \times n$ rectangles, for odd n, cannot be tiled with the L-tromino.

We are almost done. The remaining case is $3 \times n$ rectangles for n odd and at least five. These can be dealt with with a single claim.

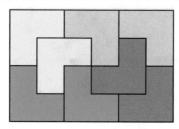

Figure 6.10: This is a L-tromino tiling of a 4 × 2 rectangle that does not have any 2 × 3 or 3 × 2 rectangle in it.

Claim 6.3   All $3k \times n$ rectangles for odd $n \geq 5$ can be tiled with the L-tromino.

Proof: Figure 6.8 shows a 5 × 9 rectangle tiled with the L-tromino. We can also find a tiling of a 5 × 6 rectangle with the L-tromino as follows.

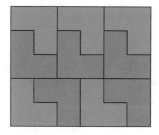

Noting that every multiple of 3 that is at least 6 is either a multiple of 6 or a multiple of 6 plus 9. The means that by concatenating the 5 × 6 and 5 × 9 rectangles give us all rectangles of size 5 × 3k. Concatenating the 3 × 2 rectangle given in Figure 6.9 let's us create a 2 × 3k rectangle that can be added to the top of the 5 × 3k rectangle to extend it by any multiple of 2 yielding rectangles of size any odd by any multiple of three larger than three. □.

To summarize the result for rectangles that can be tiled with a L-tromino, we give the following theorem, already proved, that summarizes our results.

Theorem 6.4   *The rectangles can be tiled with the L-tromino are:*

- *Any rectangle with a side length that is even and the other side length that is a multiple of three.*

- *Any rectangle with one size of odd length of five or more and the other side of length that is a multiple of three that is at least six.*

One thing that stands out in the proof of Theorem 6.4 is that the 2 × 3 or 3 × 2 rectangle appears all over the place in the proof. This gives us the natural set of problems, tile a rectangle while avoiding these configurations. An example of this appears in Figure 6.10.

Figure 6.11: The cross pentominoe.

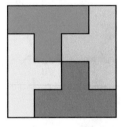

Figure 6.12: The smallest rectangle that can be tiled with a T-quadromino.

We have worked through the details of tiling rectangles with the L-tromino and this generates a large family of problems: what about other polyominos? We conclude the section with some comments on other polyominos.

Figure 6.11 shows a pentomino that is easy to solve the problem of finding the rectangles that can be tiled with it. No rectangle can be tiled with this pentomino. Figure 6.12 shows a started set for the T-quadromino, which implies any $4n \times 4m$ rectangle can be tiled with the T-quadromino. Experimenting with other polyominos is a rich source of problems.

6.2.1 SOLUTION TECHNIQUES

Most of the solution techniques are exemplified in completely solving the problem of when we can tile a rectangle with the L-tromino. That said, it is worth listing the techniques explicitly. These techniques are for solving the problem "what are all the rectangles that you can tile with this polyomino" questions. If you are simply posing questions about particular rectangles, these techniques are a superset of what you need.

- Start with the fact that the area of the object being tiled must be a multiple of the area of the polyomino in question.

- For polyominos like the one in Figure 6.11, the inability of the polyomino to form edges mean it cannot tile rectangles—be quick to notice these. Also note that you can tile shapes other than rectangles with these shapes. The cross-pentomino can be used, for example, to tile the entire plane.

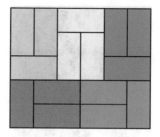

Figure 6.13: A non-uniform tiling of a 5 × 6 rectangle with copies of a 3 × 2 rectangle made of dominoes.

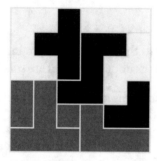

Figure 6.14: A solution to the Ten Yen puzzle.

- Once you tile a rectangle with a polyomino, that rectangle can itself be used to generate a lot of tilings. Not only rectangles that are multiples of the side lengths of the original rectangle, but also mixed use of the rectangle in different orientations. An example of this appears in Figure 6.13.

- As you pick off sizes of rectangles, look at what sizes are left to find your next target. Remember, as with 3 × n for the L-tromino, some rectangles may be impossible.

6.3   PUZZLES WITH MULTIPLE POLYOMINO SHAPES

In this section we go in another direction with polyominos: puzzles with many different polyominos in them. The puzzle in Figure 6.14 was offered for sale, starting in 1950, by *Multiple Products Corporation*, an organization that no longer exists. The puzzle was later re-issues by Kadon Enterprises.[1] This puzzle is the basis of its own problem factory. The basic puzzle is to simply put the ten pieces into the square, which can be done in 17,995 ways, a fact computed by Kadon Enterprises.

[1]http://www.gamepuzzles.com/

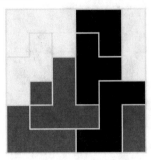

Figure 6.15: A solution to the Ten Yen puzzle in which the black pieces form a connected region, but the other colors do not.

By adding some constraints to the solutions we are interested in, we can generate a large number of problems. Here are a few examples, some of are solutions to the constrained problems.

1. Can you find 20 (or some other impressive number) solutions that are different from one another?

2. Is there a solution in which only the red (or white, or black) pieces form a connected region—some but not all of the piece of other colors may touch one another (see Figure 6.15).

3. Is there a solution in which the black, white, and red pieces form connected regions? (see Figure 6.16).

4. Is there a solution in which no two pieces of the same color touch one another?

5. Is there a solution, for each individual piece, in which that piece does not touch the edge of the square? (see Figure 6.16 for solutions to this for three of the pieces).

We have directly verified that all of these problems can be solved.

Item 5, where the goal is to solve the puzzle when a specified piece does not touch the edge of the square, has different levels of difficulty for different pieces. Getting the 1-omino away from the edge is easy. Getting the T-shaped pentomino away from the edge is a good deal harder.

Another source of a huge number of problems is to ask students to create their own polyomino puzzle in which they tile a square without re-using a polyomino shape. Three examples are shown in Figure 6.17. These are 10×10; smaller square sizes may be advisable. The steps in assigning such a problem are:

1. Teach students what polyominos are, including the idea that if one can be rotated or flipped to match another, then they are the same.

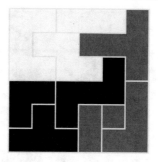

Figure 6.16: A solution to the Ten Yen puzzle in which the pieces that are the same color form connected regions.

Figure 6.17: Three examples of 10 × 10 polyomino puzzles with no repeated polyominos.

2. Use Ten Yen, or the puzzles in Figures 6.17 or 6.18, as examples of puzzles without repeated polyominos.

3. Tell students to make their own such puzzles, with the added constraint that that no one polyomino be too big.

Notice that there are trivial solutions with two large polyominos where the square is simply cut into two pieces that happen to be polyominos. You can score the students designs by computing the average size of the polyominos—this is a good way to encourage designs with many polyominos.

Many of the interesting puzzles based on Ten Yen depended on the coloring of the pieces. A second phase of having the students design their own puzzles is having them assign a small number of colors to the pieces to enable additional puzzles like the ones posed for Ten Yen.

Figure 6.18 shows 30 examples of 6 × 6 puzzles with distinct polyominos, generated with an AI-search algorithm. The AI was asked to minimize the average size of the polyominos, with the results that the smaller polyominos show up over and over. All of these puzzles are the same

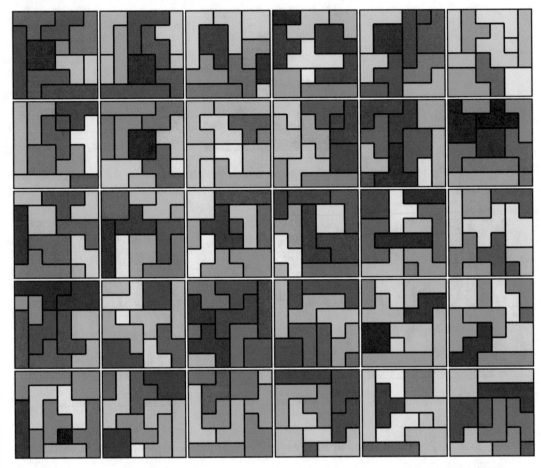

Figure 6.18: Thirty examples of 6 × 6 polyomino puzzles with no repeated polyominos.

size as the Ten Yen puzzle. These puzzles are a good starting point for coloring puzzles in a small number of colors to create puzzles based on grouping pieces by color, as in Ten Yen.

These puzzles were created in solved form, but they all have multiple solutions, in some cases thousands, like Ten Yen. This means the puzzles that ask to place a particular polyomino away from the edge are available for each of these—and probably for any of the puzzles that the students create for themselves. We have not verified that all the pieces in all of these puzzles can be placed away from the edge, but almost all can.

Making a puzzle from scratch in which all the polyominos are different from one another can be quite challenging at first—it is probably a good team effort. There are millions of these puzzles for a 6 × 6 square and an absurd number for a 10 × 10 square. Perseverance will yield solutions.

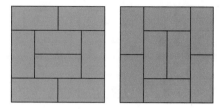

Figure 6.19: Orthogonal tilings of a 4 × 4 rectangles.

6.3.1   SOLUTION TECHNIQUES

For the problems based on Ten Yen, one valuable technique is to simply solve the puzzle over and over and harvest those solutions that have special properties, like having a particular tile away from the edge or all black tiles forming a connected group.

When the students are trying to generate their own polyomino puzzles, a good technique is to start with a grid of squares and connect squares to make polyominos. This technique was the basis of the AI that generated the examples in Figures 6.17 and 6.18. Abstract the polyominos created off the side so that spotting repeats is easy. The examples provided do not repeat polyominos, nor does Ten Yen, but repeating tiles is not intrinsically bad and is another direction that you can send your students.

Something we more-or-less skipped is coloring the non-Ten Yen puzzles in a small number of colors to set up analogs to many of the Ten Yen puzzles using the new polyomino puzzles. One example appears in the problems in the last section of the chapter. An obvious technique is to color one of these puzzles in a small number of colors to generate a solution to a puzzle, though this fixes the colors for all other potential puzzles. This suggests that coloring a puzzle solution should be done for a hard puzzle like keeping all the tiles the same color apart.

6.4   ORTHOGONAL DOMINO TILINGS

A domino tiling of a rectangle is a way of covering a rectangle with dominoes. Two such tilings of the same rectangle are *orthogonal* if, when the tilings are superimposed, each domino in one tiling covers half of two dominoes in the other tiling. An example appears in Figure 6.19. Orthogonal domino tilings (ODTs) are the key to a type of puzzle, an example of which appears in Figure 6.20. Each domino has a pair of numbers on it. The goal is to tile the rectangle with the dominoes so that each number is adjacent to an instance of itself on another polyomino.

Notice that if we look at the tiling of the solution in Figure 6.20, it is the blue tiling from Figure 6.19. If, on the other hand, we view the pairs of numbers in the solution as designating dominoes, they follow the pink tiling from Solution 6.19. This means that we can construct an orthogonal domino tiling puzzle by using a pair of orthogonal domino tilings to number the puzzle pieces. Notice also that the example in Figure 6.20 also has a repeated domino, [6 7]. If we can construct a puzzle without repeated domino labels we say the puzzle is *without repetition*.

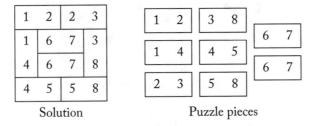

Solution   Puzzle pieces

Figure 6.20: A domino puzzle based on orthogonal tilings.

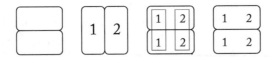

Figure 6.21: The two orthogonal domino tilings of a 2 × 2 square, turned into an ODT puzzle.

Claim 6.5   There is a pair of ODTs of size $2n \times 2n$.

Proof: Notice that the two 2×2 domino tilings, consisting of two dominoes, are orthogonal to one another. These are shown in Figure 6.21. This is a basis and we proceed by induction. Assume that there is an orthogonal $2(n-1) \times 2(n-1)$ domino tiling and examine the outer rim of a $2n \times 2n$ rectangle. Then there are two ways to fill the outer rim of the rectangle with dominoes that are, at least on the rim, orthogonal to one another. An example appears in Figure 6.22. Two orthogonal $2(n-1) \times 2(n-1)$ tilings, provided by the induction hypothesis, can be rimmed with each of the two ways of covering the rim of a $2n \times 2n$ rectangle, to yield orthogonal domino tilings of a $2n \times 2n$ rectangle.

Notice that any pair of ODTs of size $n \times k$ can be extended to a pair of ODTs of size $(n+2) \times (k+2)$. The same technique of bordering the rectangle in two ways, as in the proof of Claim 6.5, can be used to extend the tilings. □

If we have an $n \times m$ and a $k \times m$ pair of orthogonal domino tilings then we have an $(n+k) \times m$ tiling. To see this, place the m-length sides against one another for the first and second members of the two orthogonal pairs. This will place both pairs in position to demonstrate their orthogonality.

There is a pair of orthogonal domino tilings of size $2 \times 2(n+1)$ for $n \geq 1$.

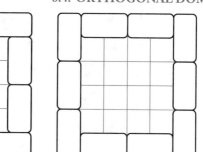

Orthogonal borders

Figure 6.22: Two distinct orthogonal borders for extending an ODT.

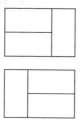

The picture above demonstrates this for $n = 1$. Extending this with the 2×2 ODTs grants us $n > 1$.

Putting all the things we have figured out so far, we have that there is a pair of ODTs of size $n \times k$, as long as $n \times k$ even.

So far it has been almost trivial to construct ODTs, but at least modest care is needed, because there are domino tilings that are not members of ODTs.

Examine the domino tiling above. An orthogonal tiling to the tiling above must have all its dominoes horizontal. Since the array is of length five, it cannot be tiled with horizontal dominoes.

6.4.1   BUILDING THE PUZZLES

Figure 6.21 suggests how to build the numbered dominoes that make up the puzzles, but let's be explicit about it now. Starting with the ODT shown in Figure 6.19, the first of the two tilings shows the physical layout of the dominoes while the second shows how to number them, like this.

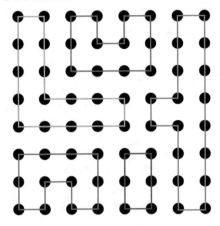

Figure 6.23: Two distinct orthogonal borders for extending an ODT.

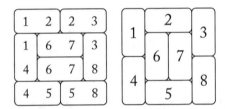

Notice that the large numbers in the second tiling are transferred to the dominoes they overlay, to make the game pieces.

Next, we make the process of finding ODTs easier. The trick used in Figure 6.22 to find two orthogonal borders can actually be used on any even length cycle by covering the cycle with dominoes in two ways. This means that if you break up a rectangle into even length cycles, then each can be tiled in two orthogonal ways giving us loads and loads of ODTs: you can chose either orientation for the first tiling for each cycle, independently. A picture of an 8 × 8 square, broken into even lengths cycles, appears in Figure 6.23. There is a cycle of length 6, two of length 12, one of length 16, and one of length 18.

Figure 6.24 shows the longest cycle in the cycle decomposition shown in Figure 6.23 covered in two ways that are orthogonal to one another. This sort of covering can be applied to each of the cycles to make an ODT. Breaking a rectangle up into even-length cycles in this fashion is easy, which means creating these puzzles is simple. To make a point more precise, the fact that each of the cycles is covered by dominoes independently of the others let's us pick either form of the cycle covering for the first tiling in the ODT. This means that, if there are n cycles in the decomposition, we can used the cycle decomposition to build 2^n different ODTs.

Once we have one puzzle designed, it turns out that we can use that puzzle to make many, many others. The simplest way is to take the number assignment step, where we lay down the

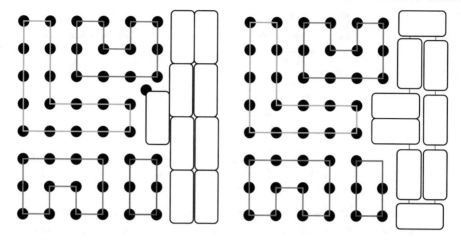

Figure 6.24: The cycle decomposition from Figure 6.23 with the longest cycle covered with dominoes in two, orthogonal fashions.

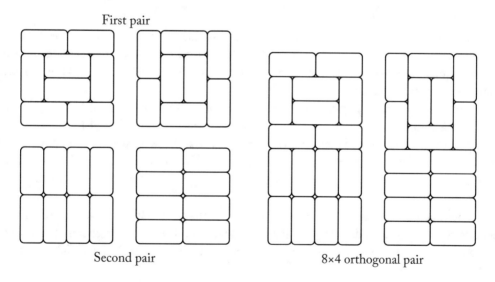

Figure 6.25: To make a new larger puzzle, two smaller puzzles can be joined.

big numbers that will turn into labels on the dominoes, and change where those numbers go. This yields a very different collection of dominoes.

There is a second method of making more puzzles once you have some puzzles: glue puzzles together. In Figure 6.25, two sets of puzzles are placed side by side to make a larger puzzle. Since we can keep joining puzzles in this fashion indefinitely, this permits the creation of puzzles of great size—probably well beyond where the puzzles are still fun to solve.

An important point about ODT is that, once we design a set of labeled dominoes that can be assembled placing all of the numbers in pairs, those pieces can be assembled in a number of ways other ways. The dominoes in the picture at the beginning of this section can also be assembled in the following fashion.

The number of different solutions that can be constructed with a given set of dominoes increases rapidly with the number of dominoes. It is important to remember that these puzzles do not have unique solutions.

6.4.2   RE-SKINNING: NON-NUMERICAL LABELS

The puzzles we have developed so far label the ends of the dominoes with numbers and each numerical label is used exactly twice, or once in the second tiling used to generate the labels. Re-using numerical labels is perfectly okay and will make the puzzle easier, or at least increase the number of possible solutions.

Numbers are perfectly good labels, but there are many other sorts of labels that can be used. Colors, animals, faces, and the other choices of labels create different flavors of puzzles. Bringing colors together would make puzzle that are more suitable for younger students. If colors are used, you can construct puzzles, by coloring sets of adjacent dominoes of the second tiling the same color, that yields a puzzle with large colored regions that can be brought together in solutions.

Labels such as animals can create another type of puzzle suitable for younger students. The directions could ask the students to put the cats, dogs, fish, birds, etc., together. These puzzles can also have other constraints like keeping the dogs and cats or the cats and fish apart in the solution to the puzzle.

6.4.3   SOLUTION TECHNIQUES

The techniques for building the puzzles in this section are actually quite easy. The basis of the building technique is to take a rectangle and divide it into cycles of even length, covering each cycle with dominoes in two orthogonal fashions. Longer cycles, and more so avoiding really short cycles, lead to more difficult puzzles. Try doodling the decompositions of rectangles into even length cycles. Note also that any connected region that can be broken into even length cycles can form the basis of a puzzle.

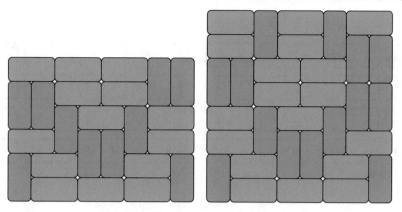

Figure 6.26: A starter set of even by even indecomposable tilings.

Once a puzzle is constructed, there are not a lot of techniques for solving the puzzles beyond trial and error. The trial and error can be made somewhat systematic. Start with one domino and then, over and over, immediately place another domino with a label that matches one, as yet unpaired, in the partial puzzle. It is easy to get stuck, at which point you backtrack one, or a few, dominoes. This method is fairly efficient at finding solutions.

6.5 EXAMPLE PROBLEMS AND PUZZLES

6.5.1 INDECOMPOSABLE AND STANDARD DOMINO TILINGS

Any even-sized rectangle can be tiled with dominoes. It turns out that any even split between horizontal and vertical dominos is possible with a rectangle that has area that is a multiple of four while rectangles of even area that is not a multiple of four can be tiles with any even-odd split of vertical and horizontal dominos. This let's you know which problems are reasonable to pose, short of demanding that the tilings also be indecomposable.

We know that an indecomposable tiling is impossible unless the rectangle is of even area, has no dimension shorter that five, and is not exactly 6×6. Using the techniques from Figure 6.7 we can start with the original 5×6 indecomposable domino tiling and construct an indecomposable even × odd tiling. The indecomposable tilings shown in Figure 6.26 can be extended by the same technique to any even × even size, meaning all the sizes that are not forbidden are possible. Remember that bigger is harder.

Figure 6.27 has four sets of dominos created with ODTs. Under each of the pictures is the size of rectangle the set of dominos were designed on: it may be possible to assemble them into other sizes of rectangle.

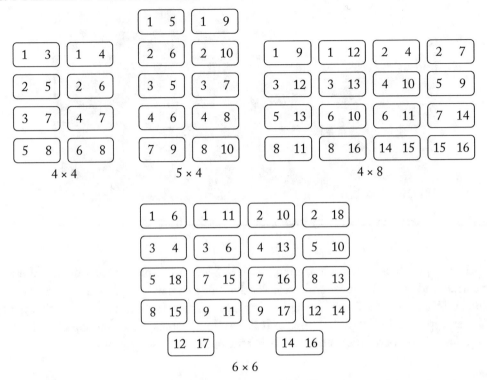

Figure 6.27: Four sets of dominos created with ODTs.

6.5.2 TILING A RECTANGLE WITH ONE POLYOMINO

Starting with the easiest case, if you have the $k \times 1$ polyomino then, as long as one dimension of a rectangle is a multiple of k, you can tile it. The rule that the area of the rectangle must be a multiple of the area of the polyomino does not quite work: consider trying to tile a 2×2 square with a 4×1 polyomino. When k is a prime number, then having an area that is a multiple of k is all that is required.

In general, when the polyomino is a rectangle, which the $k \times 1$ rectangles are an example of, the problem becomes less challenging when the polyomino is square, and is never all that hard. That is the reason that the example in which all the details were worked out used the L-tromino; it is the first moderately difficult case. The problems for tiling a rectangle with the 3-tromino are given in the chapter.

The L-tromino has the interesting property that is can be tiled with smaller copies of itself as shown in Figure 6.28. This self-tiling suggests a couple of interesting problems.

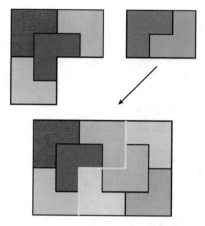

Figure 6.28: The self-tiling property of the L-tromino and an example of changing a two-tromino tiling into a eight-tromino tiling.

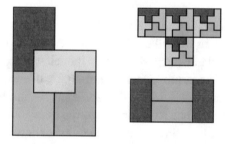

Figure 6.29: Additional examples of self-tiling polyominos.

- Since the self-tiling multiplies the number of L-trominos by four, the self-tiling can be the basis of a problem like "tile a 16×24 rectangle with trominos so that it is not made up of rectangular blocks."

- Find other polyominos that are self-tiling. All the rectangular polyominos are self tiling, which is sort of boring, as is the L-tromino. Other examples of self-tiling polyominos are given in Figure 6.29.

- It is trivial that the domino is self-tiling—but are there self tilings of the domino that can be used to expand domino tilings with special properties and preserve those special properties? An example of a trivial domino tiling that can expand an indecomposable domino tiling while preserving indecomposability appears in Figure 6.30.

There are many other self-tiling polyominos, so the question "can you find a self-tiling polyomino" is available. Probably you should exclude the L-tromino and use the Example from

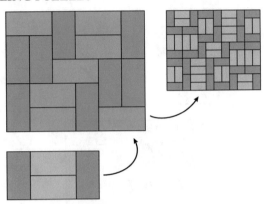

Figure 6.30: Using a simple self tiling of the domino to expand an indecomposable tiling into a much larger indecomposable tiling.

Figure 6.28 to motivate the question. Looking at the self-tiling of the T-quadromino in Figure 6.29, any polyomino that can can fill a square can then use copies of that square to build a larger copy of itself. This is a general-purpose hack for the self-tiling problem that clever students may discover for themselves. The pentomino in Figure 6.29, like the L-tromino in Figure 6.28, uses only four copies. The added condition "using the smallest number of copies of the original polyomino" is a way to an an elegance requirement to the question.

Figure 6.30 shows an application of one of the more-or-less trivial self tilings, a self tiling of the domino. It turns out that is we use the self tiling to expand the original example of an indecomposable 5×6 domino tiling then we get an indecomposable 10×12 domino tiling. This is a much faster way of creating bigger indecomposable domino tilings than the one that bordered an indecomposable tiling with a thin rectangle and then restored indecomposability by rotating a 2×2 sub-square tiled with dominos to restore indecomposability. Asking students to perform this construction and then verify indecomposability, followed by the question "will it work if we do it again?" is a good, if somewhat large, question.

Notice also that it is possible to tile the shape of one polyomino with another. This let's you expand a rectangle tiled with one polyomino to a much larger rectangle tiled with a different polyomino. This process of decomposing one polyomino shape into copies of another is a type of product or multiplication on polyominos. Suggest this possibility to students and see what they can do with it.

We conclude with Figure 6.31 which shows a number of polyominos that can tile a rectangle. The question for these is either "what is the smallest rectangle that can be tiled with this shape" or "can you tile a $k \times m$ rectangle with this shape." For the latter, make sure there is an answer available before assigning the problem. Notice that a rectangle tiles a rectangle just with one copy of the polyomino sitting there being itself.

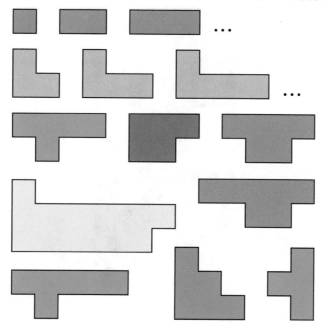

Figure 6.31: Here are examples of polyominos that can tile rectangles.

Figure 6.31 includes two infinite families, the $k \times 1$ rectangles and the L-shaped polyominos of various lengths. The other polyominos are taken from the literature and are known to be able to tile some rectangles. A very hard question, essentially a project, is to determine what rectangles can be tiles, all of them, for a given polyomino, as we did with the L-tromino in the chapter.

6.5.3   TILINGS WITH MANY POLYOMINO SHAPES

Many of the problems poses in the chapter for the Ten Yen puzzle were not solved in the chapter—the authors verified that they are solvable. Figure 6.32 shows one of the polyomino puzzles that appears in Figure 6.18, re-colored so that the various puzzles given for Ten-Yen can be worked again with a different collection of polyominos. The solubility of these puzzles has not been verified, making this more of an applied problem for your students, though the two solutions given verify five of the possible puzzles in which a piece is placed away from the edge and the "all ivory pieces together" puzzle.

When solving the puzzle in Figure 6.32, and to a degree when solving the original Ten Yen puzzle, a good technique is to place larger and more irregular pieces first. This is even more helpful with the next problem that we pose: filling enlarged versions of the five possible tetrominos with the Ten Yen pieces. A tetromino has four squares; replace each with a 3×3 grid to

Figure 6.32: Two solutions to a non-Ten Yen puzzle with three colors of polyominos.

create a shape with area 36. Figure 6.33 contains a solution for each of the five tetrominos. One of them is the original Tetris puzzle.

The solutions shown in Figure 6.33 are for a natural collection of shapes of area 36, but it also opens the question "what other shapes of size 36 can be made with the Ten Yen pieces?" The alternate puzzle given in Figure 6.32 can also be used to fill other shapes. The directions for constructing your own polyomino puzzles can also be applied to making puzzles with shapes other than a square.

Here is a suggestion for a longer term project. Polyomino pieces can be made by gluing together small wooden cubes, available at craft stores. Without using glue, used the cubes to make a person's initials. This gives you shapes for the initials. Now design polyominos that fill that shape and apply glue to make the polyominos. Finally, paint the polyominos. It might be interesting to use shades of a different color for each of the letters. In you are feeling ambitious, you might go as far as making a whole name.

This chapter gives dozens of possible polyomino puzzle problems—which means it missed almost all of them. Constructing new puzzles is one of the most creative sorts of activity you can assign, keeping in mind that the popularity will vary with different students.

Figure 6.33: Solutions to tiling enlarged tetrominos with the pieces from Ten Yen.

CHAPTER 7

Problems-Based on Graph Theory

7.1 WHAT IS GRAPH THEORY?

This chapter gives a number of different problems that are based on graph theory. This section is a brief introduction to graph theory, in case you are not already familiar with the area. Graph theory is a very simple area of mathematics.

A graph has two parts—a collection of objects, called *vertices*, and a collection of *edges* that represent an association between pairs of objects. The edges are drawn from one object to another. Figure 7.1 shows a graph with six vertices and nine edges, meaning that nine of the possible pairs of objects are associated in some way.

There are lots of versions of graphs. The one we are using does not allow an object to be associated with itself—and edge cannot be a loop from an object to that same object. The pairs that are associated by edges do not have any sort of directionality—the pair of objects is just connected.

Graphs are useful for giving a very simple way of defining a pattern of relationships on a set of objects. The number of edges ending at a vertex is called the *degree* or *contact number* of that vertex. All the objects in the graph shown in Figure 7.1 have a degree (or contact number) of three.

A graph is defined by its vertices and edges; Figure 7.1 shows a *drawing* of the graph. A graph potentially has a number of different drawings. One of the pairs of edges in Figure 7.1 cross one another—shown by a gap in the edge that goes behind the other. There is another drawing of that graph that does not have any edges crossing. If any of the drawings can be drawn with no pairs of edges crossing then we say the graph is *planar*—meaning it can be drawn in the plane without any edge crossings. Section 7.4 is based on the fact that some graphs are not planar.

7.2 CAN YOU DRAW THIS WITHOUT LIFTING YOUR PENCIL?

Figure 7.2 displays three line drawings. Two of these can be drawn without lifting the drawing implement, one cannot. This question—can this drawing be reproduced without lifting your pencil and not go over any line already drawn?—is the basis of a problem factory.

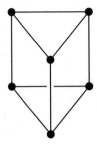

Figure 7.1: An example of a graph with six vertices and nine edges.

Figure 7.2: Three line drawings.

The key to this problem factory is a theorem that tells us if a shape can be drawn without lifting the pencil. The theory of *Euler cycles* provides that theorem. At each place where more than two lines come together, determine if the number of lines coming together is even or odd. A shape with no odd junctions can be drawn without lifting your pencil, or whatever drawing implement is being used. A shape with exactly two odd junctions can also be drawn—but the drawing process must begin and end at the odd junctions. A shape with more than two odd junctions cannot be drawn without lifting the pencil off of the page.

Examining the shapes in Figure 7.2 we see that all the junctions of the first shape are meetings of 4 lines and so the shape can be drawn; the last also has only meetings of 4 lines and so can be drawn. The second shape has *six* junctions where three lines meet and so cannot be drawn without lifting the pencil.

Let's key the first and third picture in Figure 7.2. For the first figure, start at any vertex of the hexagon and draw the triangle that touches there. Then draw one edge of the hexagon, in either direction. This will put you at a vertex of the second triangle: draw it. Now complete the hexagon. For the third picture, start at the intersection of the inner circle with one of the others. Draw the outer circle; draw an arc of the inner circle to arrive at an intersection with the next outer circle; draw that. Continue in like fashion until all the outer circles are drawn and then complete the inner circle.

As with a number of the other problem factories that generate possible and impossible problems, the impossible problems can be saved by modifying the question to be *what is the smallest number of times the drawing implement must be lifted?* An easy way to construct puzzles

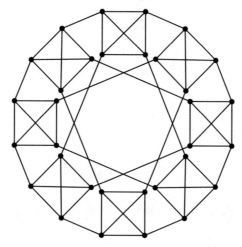

Figure 7.3: An extraordinarily symmetric and regular contact network showing the pattern of contacts between 32 people, each of which has contact with exactly 4 other people.

that can be drawn is to draw several closed curves, including polygons, that intersect one another. Adding arcs to this that begin and end at existing points in the drawing can create odd intersections and let you create pictures that can't be drawn.

7.3   ARE THESE THE NEIGHBOR NUMBERS FOR A CONTACT NETWORK?

A graph that is used to describe contacts between people is called a *contact network*. A *contact* can be any sort of connection between two people, such as two people going to the same school, or working at the same place, or traveling on the same bus, and so on. In pure math we tend to call networks *graphs*; when we are working on things like contacts between people that drive epidemics or trying to trace gossip between people, we call them *networks*. Multiple names for what are essentially the same object arise from people discovering the same cool sorts of mathematical structures independently for different purposes. The different names are actually somewhat useful, as they call to mind the applications that use the name.

   In this section, we look into the question of the pattern of numbers of neighbors in a contact network. The picture in Figure 7.3 represents a contact network with 32 people (the dots) each of which is in contact with four other people, as shown by the lines in the network. The problems we will look at in this post are of the following form. Given a sequence of numbers, can those numbers be the numbers of neighbors in a contact network? The picture at the top of post shows that 4, 4, 4, . . ., 4 (32 fours) *can* be the number of neighbors in a contact network. We call these sequences the *contact numbers* of the network. These questions make a good problem

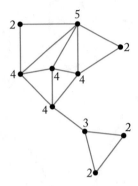

Figure 7.4: A contact network with the number of neighbors for each individual in the network annotated in green.

factory because there are many sequences that are the numbers of neighbors in a contact network, but there are also many that are not, making the questions challenging and real.

Let's start by looking at the sort of questions posed by this problem factory: "is the sequence 2, 2, 2, 2, 3, 4, 4, 4, 4, 5 a possible sequence of numbers of neighbors for a contact network of ten people?" The picture in Figure 7.4 answers the question in the affirmative. The dots represent people in the network, the lines represent interpersonal contacts, and the green number gives the number of neighbors each person has. Since there are many sequences of positive numbers that are the numbers of neighbors in a contact network, this sort of problem makes a really large problem factory. Most sequences are not the neighbor counts of a contact network. In fact there are a number of rules for this kind of sequence.

- The numbers cannot be negative—there is no such thing as having less than zero neighbors. A zero means an isolated dot in the network, i.e., a person who has no contacts.

- The numbers can be, at most, one less than the number of people (or dots) in the contact network.

- The number of people with an odd number of neighbors in a sequence must be even.

- In a sequence, there must be at least one repeated number.

Sequences that can be the numbers of neighbors in a contact network are said to be *graphical sequences*. Violating the rules above, except for the first, which is obvious, is an easy way to make a sequence that is not the numbers of neighbors in a contact network, forming a *non-graphical sequence*. Proving that these rules are true is not too hard and makes a good advanced problem for a sharp student. These rules make it easy to give non-graphical sequences. The third rule tells us, for example, that 2, 2, 2, 2, 5, 5, 5, 5, 5, 5, 5 is not graphical—it has seven odd

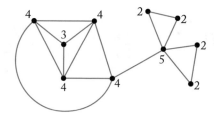

Figure 7.5: A contact network with the number of neighbors for each individual in the network annotated in green.

numbers in it. The fourth rule tells us that 1, 2, 3, 4, 5, 6, 7, 8 is not a graphical sequence, because it has no repeated numbers.

7.3.1 GENERATING GRAPHICAL SEQUENCE PROBLEMS

In general, if one of the rules given above does not tell you a sequence is not graphical, then it is pretty hard to tell if a sequence is graphical or not. The simple way to get a graphical sequence is this:

1. sketch a network and

2. count the number of neighbors of each person.

Your sketch is what mathematicians call a *witness* that the sequence is graphical. If you want every person to have the same number of neighbors then we have any sequence made of a single even number or any sequence of even length made of a single odd number is graphical, as long as the sequence obeys the first two rules.

An important fact about these problems is this. Only a very few of the contact network problems that have answers at all have unique answers. Mostly the ones with unique answers are really small ones like 2, 2, 2 which is uniquely solved by a network that looks like a triangle. Figure 7.5 gives a second network for the sequence 2, 2, 2, 2, 3, 4, 4, 4, 4, 5, the first of which appear in Figure 7.4. In fact, if you want to make these problems into longer exercises, ask students or teams of students to find as many different solutions as they can. This is a little tricky because it is possible to draw the same network in different ways. The two examples for the sequence 2, 2, 2, 2, 3, 4, 4, 4, 4, 5 are different because the triangle with a single connection back to the rest of the network only exists in the first example, not in the second—be sure to ask you students to find these distinguishing marks.

7.3.2 SOME CONVENIENT NOTATION FOR SPECIFYING SEQUENCES

There is a way to avoid writing really long sequences, as long as there are lots of repetitions of some of the numbers appearing in the sequence. Write each number in the sequence once and

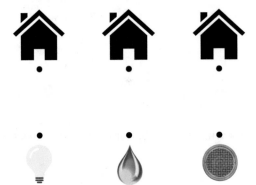

Figure 7.6: The utility problem: can three houses be connected to water, electricity, and sewage without any of the connections crossing one another?

give the number of times it appears in the sequence as an exponent so the sequence for our simple example 2, 2, 2, 2, 3, 4, 4, 4, 4, 5 becomes $2^4 3^1 4^4 5^1$. The 32-person network at the top of the post is written 4^{32}, because it has 32 people with 4 neighbors.

There is no requirement that the networks be connected—they can have different pieces that are not connected to one another. In case one of your students thinks of this, a person may not be their own neighbor. When your students are working these problems, note that many of these networks cannot be drawn unless some of the connections cross over one another. A nice contest built on this problem factory is to have teams of students generate problems and then, for each problem, have all the other teams work the problem as a speed trial.

The question "Is this sequence graphical" does not, in general, have a simple solution unless one of the rules invalidates it. There are lots of special purpose solutions. There are also some rules (hundreds, actually) beyond the four we give, but these involve fairly advanced mathematics. In spite of this, it is worth asking your students if they see any additional rules.

7.4 PUZZLES EXPLOITING PLANARITY OR THE LACK THEREOF

Recall from the discussion in Chapter 1 the utility hookup problem (see Figure 7.6).

If we remove the "houses and the utilities" skin of this problem, what we are trying to do is draw some lines or curves, called **edges**, from one group of three objects to another group of three objects. We can make the problem a little more abstract by treating each object as a dot, called a **vertex** (plural **vertices**). Give ordinate labels to the vertices in group 1 and group 2, so that the vertex 1, vertex 2, and vertex 3 are in group 1, and vertex 4, vertex 5, and vertex 6 are in group 2. Now we want it so that vertex 1 from group 1 is connected to each vertex in group 2 by drawing single edge going from vertex 1 to vertex 4, another edge from vertex 1 to vertex 5, and another edge going from vertex 1 to vertex 6, with none of those edges crossing themselves or

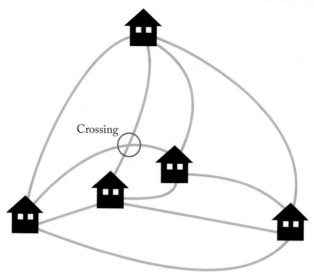

Figure 7.7: The arrangement of secret passages for the resistance with the fewest crossings.

each other. Then, we want to connect vertex 2 in a similar fashion to vertices 4, 5, and 6, and if you draw it out you can see that we are able to do it without any edges crossing. However, when you try to connect vertex 3 to vertices 4, 5, and 6, you'll notice that you must cross edges once in order to complete the picture, which we call a **graph**. The problem of drawing **graphs**, which are a collection of **vertices** and **edges**, so that no edges cross one another, is called the **planarity problem**.

From graph theory research, we know that some graphs can be drawn without any edges crossing, they are called **planar graphs**. We also know that there are two kinds of graphs that are not planar. The first is any graph that has $K_{3,3}$, which is the graph we previously discussed as the abstraction of the utility problem, buried in it as a subgraph (meaning there might be more edges and vertices in the graph, but $K_{3,3}$ is part of the graph's structure).

The other kind of graphs are ones that have K_5 as a subgraph. K_5 is the graph with 5 vertices, drawn so that all vertices are connected to one another. The following example demonstrates how K_5 could be skinned as an problem to work on: five members of the resistance have houses with basements in which secret doors can be installed. If we want to have secret passages so that every house is directly connected to all of the other houses, try connect the houses in such a way as the passages don't intersect, since each intersection represents a possible security breach. Making one passage go under another is too hard because of rock under the city. Figure 7.7 exemplifies that it is impossible to draw the picture with no crossings.

It is possible to phrase problems about K_5 that are similar to the utility problem, like the secret passage problem, but there is another collection of problems here. Draw the graph so as

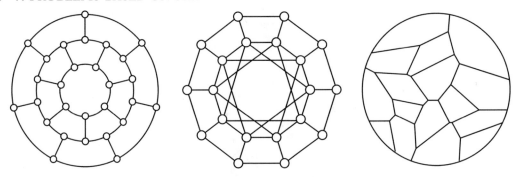

Figure 7.8: Possible boards for the Grundy puzzle.

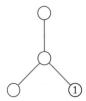

Figure 7.9: The middle vertex is adjacent to each of the outer vertices, but the outer vertices are not adjacent to one another.

to find the *minimal* number of crossing edges or intersecting passages possible. This minimum number of intersections in any drawing of the graph is called its *crossing number*.

The problem factory here is to exploit our knowledge of which graphs are planar or non-planar to pose problems that are variations on the "hook up the utilities" or "connect the re-sistance" problems. If a graph is planar then the problem is possible. If the graph is not planar then the goal could be to discover or demonstrate impossibility or to minimize the number of edge-crossings.

7.5   THE GRUNDY PUZZLE

The *Grundy puzzle* starts with a graph with large, unlabeled vertices like the one shown in Fig-ure 7.8. The first two use the dots-and-lines representation we have been using so far. The last one shows the graph as a *map*, where the regions are the vertices and two regions are neighbors if they share a border. Moves for the Grundy puzzle consist of writing a positive whole number in an empty vertex. This number may not be the same as a number previously written on another vertex connected by an edge. We say that two vertices connected by an edge are **adjacent**. For example, consider Figure 7.9.

Notice that there is a 1 placed on a vertex in Figure 7.9. That means that the vertex in the middle can't have the number 1 placed on it, which would mean a 2 or higher would have to be written on that vertex. If we wanted to continue to fill out the vertices of the graph, the other two vertices without a number could each have any number other than 2. The solution "a two in the middle and a one everywhere else" yields the smallest possible total of five.

Once you have assigned numbers to all of the vertices on a graph, your score is the sum of all of the numbers on the vertices. There are four goals for problem factories based on this mechanic:

1. maximize the score,

2. minimize the score,

3. maximize the largest number used, and

4. minimize the largest number used.

The instructor can present a graph with some of the vertices are already numbered, making the problem more or less challenging. Another interesting puzzle factory for more advanced students would be to come up with a graph that has the property of having a specific maximum or minimum number that can be used no matter how the puzzle is filled out.

A variation of the Grundy puzzle would be to use a map, rather than a graph, and write the numbers in countries. This is the same puzzle; using a map simply employs an alternate representation for the graph and forces the graph to be planar. A student writes numbers in the countries and countries that share a border may not have the same number in them. The third board in Figure 7.8 shows and example of a board of the Grundy puzzle, re-skinned to use a map-like representation.

The Grundy puzzle board that appears third in Figure 7.8 has some advantages over the other two example graphs. The other graphs have exactly three neighbors for each vertex. Some of the countries on the map-style board have six or seven neighbors. This creates the potential for (or risk of) much larger numbers being used during play.

We end the section by noting that the Grundy puzzle can be turned into a competitive game. Two players alternate filling in numbers, with different colors of ink or pencil. Each player's score is the sum of the numbers written in their own color with the winner being the player with the smallest score. Note that the graph needs to have an even number of vertices so that each player makes the same number of moves. This creates a good deal of strategy in trying to force the other player to take large numbers.

7.6   EDGE–SUM PUZZLES

An *edge-sum puzzle* starts with a graph. The student assigns positive whole numbers to the vertices so that, if we label the edges with the sum of the numbers assigned to the vertices the edge connects, then all of the numbers on vertices and edges are unique. The score for an edge sum puzzle is the sum of all the numbers used. The goal is to find the lowest possible score. Figure 7.10

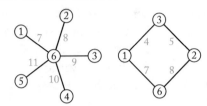

Figure 7.10: An example of optimal solutions to edge-sum puzzles. The left graph scores 66, the right one scores 36.

shows a best-possible solution to two edge-sum puzzles. These puzzles are a generalization of Peter Taylor's *Bovine Math* puzzles.[1]

This kind of puzzle is a wonderful problem factory, since there aren't really any restrictions on the kind of graphs you can use. Here are some things we know to help you decide what kind of graphs to use when creating a puzzle.

A *star graph* is a graph that has one central vertex that is connected to every other vertex in the graph, while every other vertex is only shares an edge with the central vertex. Consider the graph on the left side in Figure 7.10. This is an example of a star graph with 6 vertices, and the vertex labeled 6 would be the central vertex.

The star graph creates a fairly easy puzzle, since it has an optimal solution as long as you label the central vertex with the total number of vertices in the puzzle and then put the smaller number on the other vertices.

A more difficult kind of puzzle is one based on a graph called the *complete bipartite graph*. Complete bipartite graphs are characterized by having two groups of vertices. Each vertex in group 1 shares an edge with every vertex in group 2, but shares no edges with any other vertex in group 1. Similarly, each vertex in group 2 shares an edge with every vertex in group 1, but shares no edges with any other vertex in group 2. An example of a complete bipartite graph with 2 groups of 3 vertices each is given in Figure 7.11. General notation for a bipartite graph is $K_{n,m}$, where n is the number of vertices in the first group and m is the number of vertices in the second group.

Based on some mathematics we've done, we know that every complete bipartite graph has a perfect solution as long as you do the following: for some $K_{n,m}$ pick the larger group of n vertices (if it is the same number, then just pick the first group), and label the vertices 1 through n. Then label the vertices in the second group of vertices with the numbers $n + 1$, $2n + 2$, $3n + 3$, and so on. For example, in Figure 7.11, $n = 3$, so the vertices in the left group are labeled 1, 2, and 3, and the vertices in the right group are labeled $3 + 1 = 4$, $2 \cdot 3 + 2 = 8$, and $3 \cdot 3 + 3 = 12$.

[1]http://www.bovinemath.com/

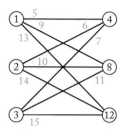

Figure 7.11: A perfect solution to the edge-sum puzzle for the graph $K_{3,3}$, the complete bipartite graph on 3 and 3 vertices. The score for this solution is 120.

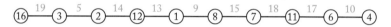

Figure 7.12: Example of the 10-path.

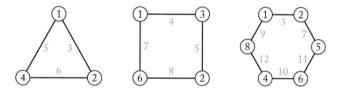

Figure 7.13: Solutions to the 3, 4, and 6 cycles.

A *path* is a graph where a group of vertices form a chain, so that vertex 1 shares an edge with vertex 2, and vertex 2 shares an edge with vertex 3, and so on. If there are 10 vertices in a path, for example, vertex 1 and vertex 10 each only have one edge attached to them, but every other vertex has 2 edges attached to it, as shown in Figure 7.12.

We know that perfect solutions exist for paths of length 2–12, but unlike the star or bipartite graphs, we are uncertain about a general rule for finding those perfect solutions.

A *cycle* graph is a like a chain, with the exception that the first and last vertex share an edge.

So far, we know that cycles of length 3, 4, 6, 7, 9, 10, and 12 shown in Figures 7.13, 7.14, 7.15 and have perfect solutions, and we've been able to show that cycles of length 5, 8, and 11 do not have perfect solutions.

At this point we hope that we've inspired you to come up with your own edge-sum puzzles. These make great problem factories because all you need are some open circles and some edges connecting them and you can generate an incredible variety of problems that share the same characteristics but fundamentally differ in their solutions. It is a near infinite test bed from which you can create new problems very quickly.

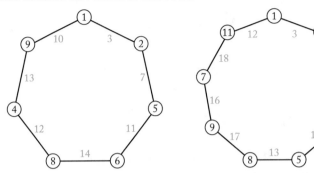

Figure 7.14: Solutions to the 7 and 9 cycles.

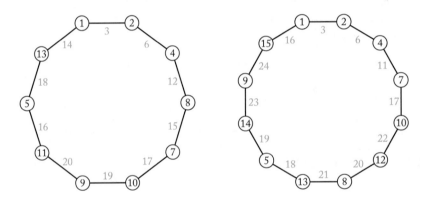

Figure 7.15: Solutions to the 10 and 12 cycles.

7.7   TRAVELING SALESMAN PROBLEMS

The Traveling Salesman Problem is famous for being very easy to understand and very difficult to create a general algorithm for finding optimal solutions. The basic version of the traveling salesman problem can be phrased as "given a list of cities and the distances between each pair of cities, what is the shortest possible route that visits each city exactly once and returns to the starting city?" The traveling salesman problem is a fantastic problem factory because you can generate traveling salesman problems using your own local geography. Pick 5–10 towns or cities near your own, and have your students find roads connecting the those places and the distances between all of them. It is also fairly quick to create a list of cities and distances on the fly. Have the students draw the map, where the cities are vertices and the edges connecting the vertices are labeled with the distances along the roads. Once the map is completed, students can come up with the best solution that they can. Having students comment on other solutions and compare them with their own is a great way to encourage positive feedback and student revisions to their

Table 7.1: Distance between cities

Distance	A	B	C	D	E
A	0	10.77032961	6.32455532	5	8.544003745
B	10.77032961	0	7.211102551	6.08276253	8.062257748
C	6.32455532	7.211102551	0	5.385164807	10.44030651
D	5	6.08276253	5.385164807	0	5.099019514
E	8.544003745	8.062257748	10.44030651	5.099019514	0

own work. An example Traveling Salesman Problem follows. Here is a list of cities and their locations.

- A is located at position (11, 1).

- B is located at position (7, 15).

- C is located at position (19, 13).

- D is located at position (19, 14).

- E is located at position (3, 8).

We've included the distances between the cities as part of the problem, but a nice extension to this problem factory is having the students find the distances (see Table 7.1). Or even better, if your students are good at coding, have them write a program to calculate the intercity distances quickly. Notice that the table is symmetric along the left to right diagonal axis of the table, since the distance from city A to city B is the same as the distance from city B to city A.

Often looking at a visual of the cities is helpful when thinking about creating an optimal route for the traveling salesman problem. We've included a map of the locations of the cities as part of this problem in Figure 7.16.

It can also be useful to draw a map with the connections between the cities as well, as shown in Figure 7.17.

The solution to this example of the traveling salesman problem is given below.

Coordinates: A(11,1), B(7,15), C(13,13), D(8,14), E(3,8)
City order: A C D B E
Length of minimal tour: 37.4

7.8   EXAMPLE PROBLEMS AND PUZZLES

We do not include examples of edge-sum puzzles in this section because there are numerous examples in Section 7.6.

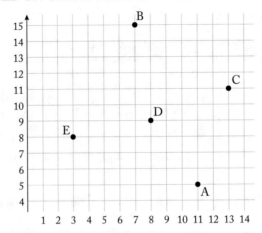

Figure 7.16: A map of the cities and their relative locations.

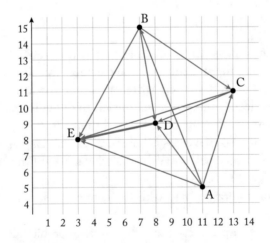

Figure 7.17: A map of the cities and their relative locations.

Which of the following shapes can be drawn without lifting your pencil? Draw the ones that are possible.

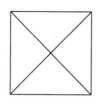
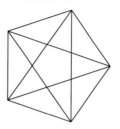

Answer: The first, third, fourth, and fifth shapes can be drawn without lifting your pencil. Find the sequence of contact numbers for the following graph.

Answer: $6^2 5^1 4^5 3^3$.

The following sequences are all graphical—find a network that is a witness.

1. $2^6 6^2$

2. $2^7 7^2$

3. $3^4 4^3$

4. $3^2 5^6$

5. $3^5 7^3$

6. $4^{10} 5^2$

7. $3^9 5^1 6^2$

8. $1^1 3^5 4^2 6^3$

9. $2^2 3^2 4^3 5^2$

10. $1^1 2^3 3^1 4^2 5^2$

Find a non-graphical sequence that does not violate the first two of the four rules given in Section 7.3.

Here are six possible graphs for the Grundy game.

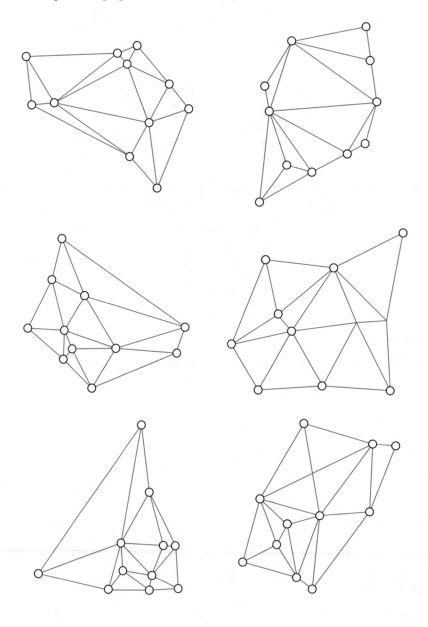

Here are six possible maps for the Grundy game.

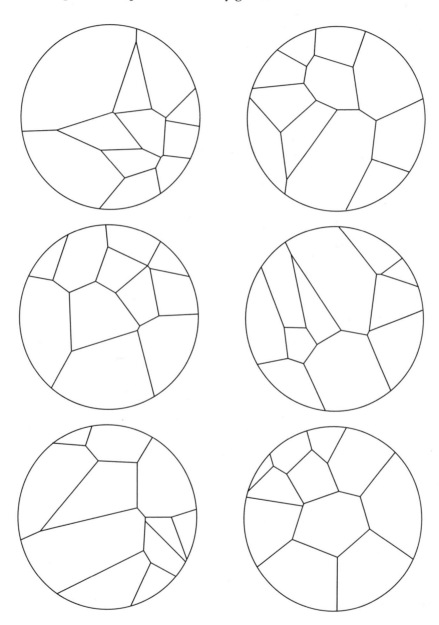

The next five problems are on the traveling salesman problem. Each problem has three different TSP puzzles. The goal is to find the order on the cities that makes the TSP tour as short as possible. The solution can be a distance or a sketch of the order of the cities. Sketches of the solutions appear after each problem.

1. Find the shortest tour.

   Coordinates: (91,79) (79,59) (24,39) (54,40) (20,69)
   City order: 0 1 3 2 4
   Length of minimal tour: 186.7

2. Find the shortest tour.

   Coordinates: (6,82) (21,36) (42,32) (53,26) (46,65)
   City order: 0 1 2 3 4
   Length of minimal tour: 165.4

3. Find the shortest tour.

   Coordinates: (73,33) (91,49) (58,82) (20,56) (68,89)
   City order: 0 1 4 2 3
   Length of minimal tour: 186.2

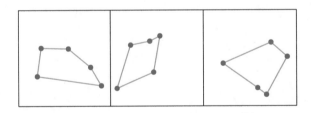

1. Find the shortest tour.

   Coordinates: (89,33) (85,66) (7,93) (6,25) (21,51) (37,94)
   City order: 0 1 5 2 4 3
   Length of minimal tour: 276.5

2. Find the shortest tour.

   Coordinates: (46,75) (27,53) (30,38) (67,80) (51,34) (74,92)
   City order: 0 1 2 4 3 5
   Length of minimal tour: 161.1

3. Find the shortest tour.

   Coordinates: (66,72) (77,12) (45,83) (41,33) (95,75) (18,8)
   City order: 0 2 3 5 1 4
   Length of minimal tour: 261.6

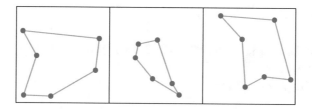

1. Find the shortest tour.

   Coordinates: (36,48) (38,69) (9,49) (62,80) (32,36) (19,84) (49,43)
   City order: 0 4 2 5 1 3 6
   Length of minimal tour: 179.2

2. Find the shortest tour.

   Coordinates: (55,10) (65,95) (40,55) (33,69) (42,83) (37,12) (6,52)
   City order: 0 2 1 4 3 6 5
   Length of minimal tour: 237.8

3. Find the shortest tour.

   Coordinates: (60,70) (21,33) (25,87) (63,56) (33,41) (83,45) (70,32)
   City order: 0 3 5 6 4 1 2
   Length of minimal tour: 201.1

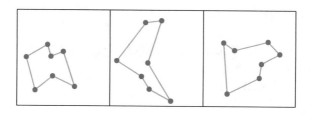

1. Find the shortest tour.

   Coordinates: (38,69) (9,49) (62,80) (32,36) (19,84) (49,43) (55,10) (65,95)
   City order: 0 7 2 5 6 3 1 4
   Length of minimal tour: 247.3

2. Find the shortest tour.

   Coordinates: (40,55) (33,69) (42,83) (37,12) (6,52) (60,70) (21,33) (25,87)
   City order: 0 3 6 4 1 7 2 5
   Length of minimal tour: 210.0

3. Find the shortest tour.

   Coordinates: (32,23) (34,36) (29,66) (80,86) (78,51) (15,92) (66,27) (50,17)
   City order: 0 1 2 5 3 4 6 7
   Length of minimal tour: 238.1

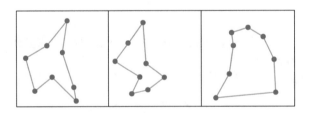

1. Find the shortest tour.

   Coordinates: (9,49) (62,80) (32,36) (19,84) (49,43) (55,10) (65,95) (40,55) (33,69)
   City order: 0 3 6 1 8 7 4 5 2
   Length of minimal tour: 255.3

2. Find the shortest tour.

   Coordinates: (42,83) (37,12) (6,52) (60,70) (21,33) (25,87) (63,56) (33,41) (83,45)
   City order: 0 3 6 8 1 7 4 2 5
   Length of minimal tour: 241.2

3. Find the shortest tour.

   Coordinates: (15,92) (66,27) (50,17) (42,77) (47,58) (15,46) (47,29) (42,42) (57,66)
   City order: 0 3 8 4 1 2 6 7 5
   Length of minimal tour: 217.1

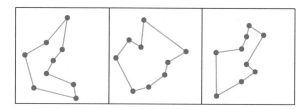

1. Find the shortest tour.

   Coordinates: (62,80) (32,36) (19,84) (49,43) (55,10)
                (65,95) (40,55) (33,69) (42,83) (37,12)
   City order: 0 3 4 9 1 6 7 2 8 5
   Length of minimal tour: 236.4

2. Find the shortest tour.

   Coordinates: (6,52) (60,70) (21,33) (25,87) (63,56)
                (33,41) (83,45) (70,32) (80,25) (26,61)
   City order: 0 2 5 7 8 6 4 1 3 9
   Length of minimal tour: 233.1

3. Find the shortest tour.

   Coordinates: (32,23) (34,36) (29,66) (80,86) (78,51)
                (15,92) (66,27) (50,17) (42,77) (47,58)
   City order: 0 1 9 2 5 8 3 4 6 7
   Length of minimal tour: 257.6

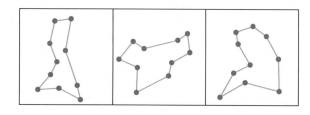

C H A P T E R   8

The Road Ahead: Other Problem Factories

In this book, we've shown you several types of problem factories. The authors have been researching problem factories for several years now and the ones chosen for this book belong to a particular category. While we used computers to generate or check our examples, all of the problem factories in this book can be worked with pencil and paper and require only very modest sorts of computer code to generate or validate them. During our research, we also found a number of problem factories that require moderately powerful AI to generate or validate the problems. This does not mean they are not problem factories, but it does put them in a different category. This chapter is going to mention a few of these more complex problem factories and discuss plans for making these available in a future book.

Chapters 2–7 give example problems or example problems and puzzles, but the chapters also explain in some detail how to construct your own problems. There is a little leakage of the problems requiring more advanced techniques for validation in this book. Edge sum puzzles, for example, have infinite families of problems that have solutions, like complete bipartite graphs. Cycles and paths, which are natural, simple examples of edge-sum puzzles, do not (yet) have obvious theorem-based solution techniques. For these we used a very simple search-based solver to provide the examples in the text. During the writing of this book, we discovered that it seems to be the case that every third cycle size, starting with five, does not have a perfect edge-sum solution. We now have a legitimately difficult math problem to work on—why do only some cycle sizes have perfect edge-sum solutions?

Problem factories, as an idea, are a simple organizing tool for a class of problems intended to make life easier for instructors. A piece of mathematical knowledge is explored and then connected to techniques for generating interesting problems and their solutions. The more complex problem factories may also provide interesting challenge problems for computer science contests. We have learned about some problem factories at conferences from colleagues, others appear in the literature, and we have some pride in creating some of them ourselves. This mix of sources strongly suggests that inviting other pedagogical researchers into the effort is a good idea and we welcome suggestions and offers of collaboration.

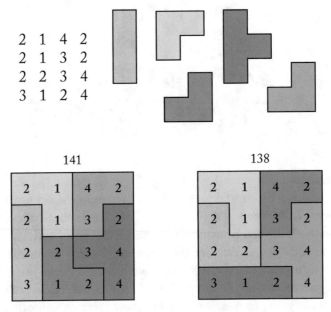

Figure 8.1: An example of a polyomino math puzzle and two solutions, with their scores.

8.1   POLYOMINO MATH PUZZLES

Figure 8.1 shows an example of a *polyomino math puzzle*. The grid of numbers and the polyominos that make up the puzzle are shown in the top half of the figure. Two solutions to the puzzle with scores are shown in the bottom half. The rules are for these puzzles are as follows.

1. The player places the polyominos on the grid of numbers.

2. The score for a polyomino is the maximum of the sum or product of the numbers under the polyomino.

3. The score for the puzzle, to be maximized, is the sum of the scores of the individual polyominos.

   Constructing these puzzles is actually a very difficult problem. First of all, a *majority* of the instances of these puzzles achieve a maximum score by **not** using some of the polyominos. This fact intruded into the initial research on the problem and was verified with a sampling study. The critical task in scoring is to cover the highest numerical values on the grid that you can with a single polyomino. This often requires a placement of a polyomino that makes covering the grid with the available polyominos impossible. Fairly powerful AI search software is required to locate instances of these puzzles where the maximum score corresponds to a placement of polyominos that covers the numerical grid.

This leads to an interesting issue–are puzzles where *not* using some polyominos yields the maximum score interesting, or are they things to be avoided? The current plan is to present both sorts (clearly marked) and let instructors evolve their own pedagogy. The polyomino math puzzles are problem factories in part because there are thousands of these problems of reasonable size that are really different from one another. Beyond that, however, one of these puzzles can be reconfigured into hundreds of these puzzles with modest effort.

Here is the trick—and you can apply it to the example in Figure 8.1. Take an optimal solution, glue the numbers to the polyominos, then lay the polyominos down to fill the grid in a different way. As we saw in Chapter 6 there are often many ways to cover a given grid with a set of polyominos. Next, un-glue the numbers from the polyominos to get a new number grid. This gives you a new number grid, and with it a new polyomino math puzzle, but there is an excellent chance it has the same optimal score. By "excellent chance" we mean that it usually does and there is a simple test to check if this is the case.

This ability to rearrange the puzzle to get a nominally different puzzle is reminiscent of techniques for rebuilding or reconfiguring the orthogonal domino tilings from Chapter 6. The diagrammatically represented linear systems in Chapter 5 have a similar ability to take a single, fundamental mathematical puzzle and present it as a large number of realized puzzles that look quite different from one another. The fact that this phenomenon has arisen several times during our investigation of problem factories makes it seem worth coining a term for this sort of problem factory. We call such puzzles *multirepresentation* problem factories.

8.2   THE TAXMAN PROBLEM

The taxman problem, also called *number shark* or *Zahlenhai* in the German literature, is a classical math puzzle with the following rules.

1. The game starts with n checks, with values of 1 through n.

2. The player picks a check and adds it to his total while the "tax man" picks up every check whose value divides that check.

3. The player may not pick a check if it has no divisors among the remaining checks.

4. Checks remaining at the end of the game also go to the tax man.

The goal of the game is for the player to maximize their score. So far, this looks like a problem we could have included in this collection since the use of different values of n does give us a problem factory and creation of instances of the problem amounts only to picking n. Notice that the fourth rule is also completely irrelevant to the goal of maximizing score. It is included because people often ask "what happens to the unused checks?"—which do not affect the player's score.

Here is the solution for $n = 12$. The player takes 11 with the taxman taking 1. The player takes 9 with the taxman taking 3. The player takes 10 with the taxman taking 5 and 2. The player takes 8 with the taxman taking 4, and finally the player takes 12 with the taxman taking 6. The remaining check 7 is ceded to the taxman. The player scores $12 + 11 + 10 + 9 + 8 = 50$, the known optimal score for $n = 12$.

It may be a bit startling that we do not have a general solution to the taxman problem. The answers are known up to $n = 701$ at the time of this writing, computed by Brian Chess,[1] but the problem of an optimal strategy for the game is, in general, unsolved. Trying to think up better heuristics for attacking the taxman problem is, itself, and interesting source of projects for sharp students.

Dan Hoey notes that:

> The Taxman game was invented by Diane Resek, a mathematician at San Francisco State University, some time in 1971–1973. She made it up as a teaching aid to motivate schoolchildren to practice arithmetic. The game was distributed for the Apple II by the Minnesota Educational Computing Corporation. It has also been distributed under the titles "Dr Factor," "Factor Blaster," and "Number Shark" ("Zahlenhai" in German). There is also an unrelated arcade-style game named Taxman that was popular on the Apple II.

Generalizing problems is a major industry in mathematics and the set of checks $\{1, 2, 3, \ldots, n\}$ is just one possibility. Leaving out some checks from the starting set creates different problems. If the player and the taxman may only take one check for a value that is repeated, the game may be generalized to multisets. Recall that multisets are like sets, but they allow repeated elements. An example of a multiset is $\{1, 1, 2, 2, 3, 4, 4, 4\}$.

Once we start proposing taxman puzzles on sets other than the numbers 1 through n and even on multisets, we want to have the power of our AI solvers available. Math teachers that make up their own problems have almost all had the experience of posing what seemed to be an entirely reasonable math problem that turned out to be much harder than they thought. With various AI solvers available, we cannot only continue the major value add of this text—providing solved problems—but also use the AI solvers to estimate how hard the problems are. If the solver has to test 72 cases to solve the problem, it is an easier one that a problem where the solver needs to test 1,324,545,221 cases.

8.3   NUMBER SENTENCE MORPHING PUZZLES

These problems are based on the ideas for number sentence puzzles in Chapter 4. Number sentences are a staple of math education in the lower grades. We propose a competitive game that starts with all the players having the same false number sentence, e.g.:

$$3 + 4 * 5 + 3 = 20.$$

[1]https://oeis.org/A019312/b019312.txt

For each player or team, a move consists of removing one symbol from the number sentence or adding one symbol to the number sentence. Here are the rules:

1. Each move consists of removing or inserting a single symbol from the number sentence. Matching left and right parenthesis are considered a single symbol for the purpose of making a move.

2. Each move must yield a well-formed (if possibly false) number sentence. So removing the $*$ from the example sentence to get

$$3 + 45 + 3 = 20$$

   is okay but removing the last 3 to get

$$2 + 4 * 5+ = 20$$

   is not a permitted move. The resulting symbols are not a properly formed number sentence.

3. Removed symbols are placed in a *symbol pool*, written to the right of the evolving number sentences.

4. Players present their moves secretly to the teacher or moderator who then puts the moves on a white board and scores them. If a move leads to a sentence that is not well formed the teacher may either decide the player loses a turn or request a new move. This choice should be consistent for all players in a given instance of the game.

5. The player(s) after each move with the smallest difference between values of the left and right side of their sentence score one point.

6. Moves should be made relatively promptly but the teacher or moderator controls the time permitted each team.

7. A player may choose to pass and not modify their number sentence.

8. A player may not pass twice in a row.

9. The game ends when one or more players achieve the minimum possible difference between left and right, which is supplied to the teacher (this is why we need our computer code).

10. At the end of the game each team that has the minimum difference scores one additional point for each team that does not.

11. At the end of the game, each team's score is reduced by the number of symbols in their symbol pool.

12. Highest score wins.

Example 8.1 *Number Sentence Morphing Play*

This is an example of number sentence morphing by one player. Notice that getting the difference between the sides of the equation to go down steadily is not practical in the example, or in most of the examples we've studied. Some planning ahead, and acceptance of large jumps in the difference, are needed to do well with number sentence morphing.

Turn	Sentence	Symbol Pool	Difference
1	$3 + 4 * 5 + 3 = 20$		6
2	$3 + 4 * 53 = 20$	$+$	195
3	$3 + 4 * 5 = 20$	$+, 3$	3
4	$3 + 4 * 5 = 203$	$+$	180
5	$3 + 4 * 5 = 20 + 3$		0

Minimum difference is zero

Number sentence morphing can be played as a solitaire in which case the player is given the minimum difference and tries to find the solution in the smallest number of moves.

There is no need for the symbol pool to start empty, another issue that we plan to explore in our next foray into problem factories. Number sentence morphing builds the following skills.

- Problem solving.

- Recognition of the difference between well formed and poorly formed number sentences.

- Order of operations.

- Developing long-term strategic thinking.

8.4   PROBLEM FACTORIES THAT PREPARE STUDENTS TO PROGRAM

Programming a computer is not exactly mathematics, but it is often math-adjacent and students that learn math gain discipline and clarity of thought that helps them program more effectively. The following is a problem factory that we constructed that is a family of puzzles that require students to create very simple programs, straddling the boundary between math and computer science.

Suppose you have two numbers x and y that start the game with a value of zero and that there are three moves you can make:

Table 8.1: Minimal computation of 23

x	y	Move
0	0	(start)
1	0	$x = 1$
1	1	$y = x$
2	1	$x \leftarrow x + y$
3	1	$x \leftarrow x + y$
4	1	$x \leftarrow x + y$
5	1	$x \leftarrow x + y$
5	5	$y = x$
1	5	$x = 1$
6	5	$x \leftarrow x + y$
11	5	$x \leftarrow x + y$
11	11	$y = x$
1	11	$x = 1$
12	11	$x \leftarrow x + y$
23	11	$x \leftarrow x + y$

1. Set x equal to 1.

2. Set y equal to x.

3. Add the value of y to x.

The puzzle is this: pick a target number N and find the smallest number of commands that can get x to have a value of N. Here is a solution for $N = 23$.

Example 8.2   A minimal computation of 23 is shown in Table 8.1.
This is a nice puzzle built around a very simple programming language with three commands. Formally, this is a register machine with a screen (x) and a single memory register (y). Since you could get x and y to both be 1 and then add y to x as many times as needed, it's clear there is a program to generate any number—but this demonstration leads to a program that is both long and less than clever.

The puzzle aspect of this problem factory is to find the *shortest* program that can generate the target number. This can lead to other programming concepts, like subroutines. We solved the "find the shortest program" problem for all $N \leq 100$ and there are some sequences of commands that appear over and over. These are the subroutines or procedures of this puzzle language. Discovering these re-usable patterns is an exercise that can prepare students for important skills like algorithm design.

8.5   WRAPPING UP THIS COLLECTION OF PROBLEM FACTORIES

Problem factories are a very natural idea within mathematics, where a set of general principles have legions of specific examples. We have also found problem factories in computer science, which is probably because one of the authors is a computer scientist. The pervasiveness of patterns in both these disciplines makes the existence of something like a problem factory certain. The presence of problem factories in other disciplines is less obvious.

Another problem factory that we will have ready for our future efforts is a polyomino math puzzle with a very limited set of polyominos where we generate the grid of numbers by rolling dice. This makes a mathematical game somewhat like the word game *Boggle*(tm). The fact that number grids often has a best solution that does not use all the polyominos is somewhat ameliorated by allowing an unlimited supply of 3-ominos of either sort to play the game—but even with this compensating design choice the best answer sometimes leaves uncovered dice. This problem factory does exist in the realm of statistics and probability, expanding the horizons a bit more.

It seems very likely that mechanics or electricity and magnetism in physics would be domains where problem factories might exist. Chemistry seems a harder place to find these because much of chemistry is extremely detail-rich, in a fashion that makes many abstractly posed problems nonsensical for all but a few values of their parameters. The question of the reach of the problem factory construct across fields beyond the purely computational is a matter for the future and another reason that we invite collaboration.

Authors' Biographies

ANDREW MCEACHERN

Andrew McEachern is an Assistant Teaching Professor in the Mathematics and Statistics Department at York University in Ontario, Canada. His research has been in the fields of DNA analysis, evolutionary computation, game theory, and education in mathematics. He has over a decade of experience in outreach mathematics at all levels. He is currently working on the problem of getting students of all ages to better understand fractions, as well as engaging with mathematics without dread. Part of his mission is to demonstrate that popular mathematics is all around us, in the form of puzzles and games, and it is his opinion that mathematics should mostly be fun, and at least a little useful. He is a huge fan of tabletop roleplaying games, which are a combination of mathematics and storytelling, his two favorite things.

DANIEL ASHLOCK

Professor Ashlock has a Doctorate in Mathematics from the California Institute of Technology as well as B.Sc. degrees in mathematics and computer science from the University of Kansas. He has taught collegiate-level mathematics for more than 40 years, winning awards for excellence in both undergradutate and graduate teaching. He has more than 300 peer-reviewed research papers including work in pure mathematics, pedagogy of mathematics, game theory, combinatorics, automatic content generation, complex systems, and bioinformatics. Problem factories are a recent interest that address a number of issues in the teaching of mathematics.

Printed in the United States
by Baker & Taylor Publisher Services